SpringerBriefs in Applied Sciences and Technology

Computational Mechanics

Series Editors

Holm Altenbach, Faculty of Mechanical Engineering,
Otto-von-Guericke-Universität Magdeburg, Magdeburg, Sachsen-Anhalt, Germany
Lucas F. M. da Silva, Department of Mechanical Engineering, Faculty of
Engineering, University of Porto, Porto, Portugal
Andreas Öchsner, Faculty of Mechanical Engineering, Esslingen University of
Applied Sciences, Esslingen, Germany

These SpringerBriefs publish concise summaries of cutting-edge research and practical applications on any subject of computational fluid dynamics, computational solid and structural mechanics, as well as multiphysics.

SpringerBriefs in Computational Mechanics are devoted to the publication of fundamentals and applications within the different classical engineering disciplines as well as in interdisciplinary fields that recently emerged between these areas.

More information about this series at http://www.springer.com/series/8886

Khameel B. Mustapha

R for Finite Element Analyses of Size-dependent Microscale Structures

 Springer

Khameel B. Mustapha
Department of Mechanical, Materials
and Manufacturing Engineering
University of Nottingham Malaysia Campus
Semenyih, Selangor, Malaysia

ISSN 2191-530X   ISSN 2191-5318   (electronic)
SpringerBriefs in Applied Sciences and Technology
ISSN 2191-5342   ISSN 2191-5350   (electronic)
SpringerBriefs in Computational Mechanics
ISBN 978-981-13-7013-7   ISBN 978-981-13-7014-4   (eBook)
https://doi.org/10.1007/978-981-13-7014-4

Library of Congress Control Number: 2019933203

This Springer imprint is published by the registered company Springer Nature Singapore Pte Ltd.
The registered company address is: 152 Beach Road, #21-01/04 Gateway East, Singapore 189721,
Singapore

Preface

The subject of this brief book is the elementary analyses of linear elastic size-dependent structures based on the modified couple stress theory. The book focuses on establishing the governing equations of size-dependent structures (restricted to beams and plates), deriving the associated finite element model and implementing the finite element models in the R programming language. The implemented functions are employed to develop a special R package called *microfiniteR*. With the R package, the book provides an interactive platform for finite element analyses of microscale beams (bending, buckling and free vibration) and plates (bending and free vibration). As far as computational tasks are concerned, the R programming language is well known for statistical computations and data analyses, but it is less associated with typical engineering computations (which it is also well equipped for). An aim of this book is to show its usefulness as a complementary tool in this regard, given the fast development of the language in recent years as a useful tool for reproducible research.

Chapter 1 of the book introduces the R programming language, beginning with the resources needed to make use of the language and ending with a list of recommended texts. In subsequent chapters (2, 3 and 4), we begin with a short introduction, move on to the requisite linear elastic model formulated via the variational method and then present the finite element model as well as the implemented R functions for the finite element analysis. In addition, the chapters are embedded with examples to demonstrate the use of the R functions by examining deformation characteristics (in the case of bending analyses) or the eigenvalues (in the case of dynamics and buckling problems). A brief summary and relevant references are provided at the end of each chapter.

Efforts have been made to check for errors. Nevertheless, the author welcomes notifications on errors or suggestions for corrections, which can be directed to KhameelB.Mustapha@nottingham.edu.my.

Semenyih, Malaysia
2019

Khameel B. Mustapha

Acknowledgements

The knowledge and skills require for writing this book have been gained from the openness of two large but seemingly disparate scientific communities: the R programming and the applied mechanics communities. I am indebted to these two communities in so many ways.

I appreciate the visionary efforts of many unsung heroes who maintained the overwhelmingly large R ecosystem. I thank the developers and contributors to the R language and the inspiring teams at RStudio for their selflessness. It would have been close to impossible to make the package developed for this book without the functionality of RStudio and the **devtools** package by Hadley Wickham and the R Core team.

With regards to the mechanics-related technical content, the information contained in this book, though brief and very elementary, arise from the seminal contributions of many excellent studies on the subject of modified couple stress theory and the finite element method. I am indebted to the authors of books and journal articles on the subject matters.

I thank the editorial and production staff of Springer Nature Singapore, and I express gratitude to Dr. Christoph Baumann for the support shown for this project.

The institutional support of the University of Nottingham Malaysia Campus is appreciated. I acknowledge the support of Malaysia's Ministry of Education (MOE) through the Fundamental Research Grant Scheme (FRGS/1/2018/TK03/UNIM/02/1).

I express gratitude to my parent and profoundly value the immense support from my wife Aminah and our kids (Hibatullah, Abdul Alim and Hameedah) in this journey.

Finally, all praise is due to the Best Writer, the Greatest and the Most Exalted.

Contents

About the Author

Khameel B. Mustapha is Assistant Professor in the Department of Mechanical, Materials and Manufacturing Engineering at the University of Nottingham (Malaysia Campus). He currently convenes *Mechanics of Solids* and *Advanced Solid Mechanics (Stress Analysis Techniques)* and he co-teaches *Computer Modelling Techniques, Additive Manufacturing and 3D Printing* and *Design, Manufacture & Project*. In a previous role as a lecturer in Swinburne University of Technology (Malaysia Campus), he convened and taught the following courses: *Solid Mechanics, Structural Mechanics, Computer Aided Engineering* and *Computer Modelling, Analysis and Visualization*.

He holds a Ph.D. in mechanical engineering from Nanyang Technological University (Singapore), B.Eng. in mechanical engineering from University of Ilorin, GCLT from Swinburne University of Technology, Australia, and PGCHE from the University of Nottingham, UK. He is Fellow of the Higher Education Academy, UK, and Member of the American Society of Mechanical Engineers. His research interest involves mechanics of composite materials; mechanics of additively manufactured components; mechanics of small-scale structures; mechanics of biological structures and energy materials; and computational methods for simulations of mechanical systems.

Chapter 1
A Brief Introduction to R, R Studio and MicrofiniteR

Abstract This chapter provides a short introduction to the R programming language and the RStudio environment to facilitate familiarity with the computer codes in the later chapters of the book. Basic instructions on relevant installations, demonstrations of some R commands, a description of **microfiniteR** (developed for this book) and a list of references for further consultations are presented.

1.1 The R Programming Language

Developed in the mid-90s at the University of Auckland by Ross Ihaka and Robert Gentleman, R is closely associated with hardcore computational statistics and it is often seen as an alternative to statistical packages such as Stata and SAS [1–3]. However, the language has emerged as a quality open-source platform for a variety of scientific programming tasks [4]. In recent years, R's appeal has spread beyond the domain of statistical computations. This is reflected in book series dedicated to spreading the knowledge of its usage across different fields [5, 6]. In the area of numerical analysis, the books by Soetaert et al. [7] and Griffiths [8] presented extensive coverages of solution techniques to a multitude of differential equations by well-tested, high-precision integrators developed in R. Different concepts on numerical quadrature, numerical differentiation, optimization and curve fitting procedures are covered in Bloomfield [9]. Schiesser [10] features the use of the language for solutions to differential equations that originate from the field of Biomedical Science and Engineering. Recently, a demonstration of the capability of the language for finite element analyses is presented by Mustapha [11].

One of the main challenges faced by many newcomers to R is its large ecosystem and the question of where to start. The good news is that there are many accessible texts on the language for beginners. In fact, some of these texts can be accessed for free via https://bookdown.org/. Besides, a gentle introduction to scientific computing in R can be found in Jones et al. [12]. For an example-driven approach to learning R, Albert and Rizzo [13], Kabacoff [14] and Teetor [15] are recommended. In-depth coverage of R can be sought in Adler [16] and Wickham [17]. Meanwhile, Allerhand [18]

© The Author(s), under exclusive license to Springer Nature Singapore Pte Ltd. 2019   1
K. B. Mustapha, *R for Finite Element Analyses of Size-dependent Microscale Structures*, SpringerBriefs in Computational Mechanics,
https://doi.org/10.1007/978-981-13-7014-4_1

provides a quick introduction to R in just 89 pages! There are also many excellent web resources, e.g.:

- https://www.statmethods.net/index.html
- https://www.r-bloggers.com/
- http://www.cookbook-r.com/Basics/.

1.2   Tools for Getting Started

1.2.1   R Installer

The first step to using the language is to obtain the R installer, which can be downloaded from the official home of the R programming language called CRAN[1]: https://cran.r-project.org/. A screenshot of the homepage is shown in Fig. 1.1. As revealed in this figure, the installer is available for different operating systems. Clicking on the link that matches your operating system will re-direct you to the relevant website. As an example, clicking on the *Download R for Windows* will re-direct you to the website for Windows installer (shown in Fig. 1.2). Follow the prompts to download, save and run the files.

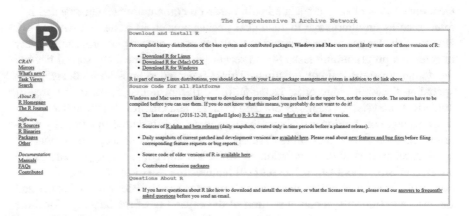

Fig. 1.1   Screenshot of a segment of the CRAN website

---

[1]CRAN stands for the Comprehensive R Archive Network.

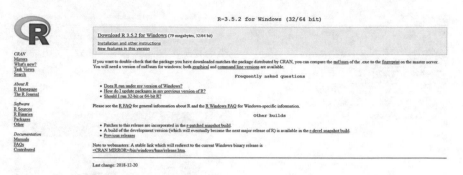

Fig. 1.2 Windows-specific website for the R installer

1.2.2 RStudio

Once R has been installed, the next step is to install RStudio which is the popular integrated development environment (IDE) for working with R. RStudio can be downloaded from: https://www.rstudio.com/products/rstudio/. After the installation, look for and click on the RStudio icon (shown in Fig. 1.3) and you should see something similar to Fig. 1.4.

Click on the small symbol pointed to by the arrow in Fig. 1.4 to reveal the full layout of the RStudio GUI as revealed in Fig. 1.5.

1.3 Overview of Basic Operations

R is a procedural and a high-level programming language like MATLAB, Basic, FORTRAN, and C++. The base R program comes with an incredible list of built-in functions and mathematical constants. It is impossible to be introduced to all of R functionalities in one go. However, a brief coverage of some of the functionalities are highlighted next.

Fig. 1.3 The RStudio icon

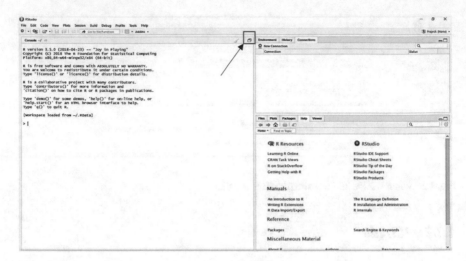

Fig. 1.4 The RStudio GUI

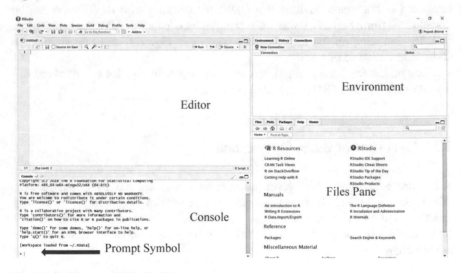

Fig. 1.5 Full layout of RStudio GUI

1.3.1 Using RStudio Console as a Calculator

The window pane named (Console) in Fig. 1.5 is typically used to display results of computations or error messages. However, it can also be used as a powerful calculator, which is a good way to start experimenting with R.

Type each of the expressions below in the console and hit the Enter key after each (ignore the part of the expression with #).

```
> # A line preceded by the "#" (without the quotes) is a comment in R
> 2*5                                    # Simple multiplication
[1] 10
> 3^4                                    # Exponentiation
[1] 81
> 3**4                                   # Another way to do exponentiation
[1] 81
> 10 + 30 - 56/3 + 90                    # Combination of operators
[1] 111.3333
> round(111.333,0)                       # Rounding the number 111.333
[1] 111
> x = 35                                 # Assigning a variable
> y = x + 100                            # Assigning and using the earlier
defined variable
> y                                      # Print the value of y
[1] 135
> (Area = pi * x^2)                      # Using a built-in variable pi
[1] 3848.451
> vec = c(20, 32, 14, 63, 91, 8)         # The command c() is used to form a
vector
> min(vec)                               # Obtain the smallest number in the
vector named vec
[1] 8
> max(vec)                               # Obtain the maximum number in vec
[1] 91
> vec[1]                                 # Obtain the first element of vec
[1] 20
> vec[1:3]                               # Obtain the first 3 elements of vec
[1] 20 32 14
> rep(vec,2)                             #  Repeat vec two times
 [1] 20 32 14 63 91  8 20 32 14 63 91  8
> vec/2                                  # Divide all elements of vec by 2
[1] 10.0 16.0  7.0 31.5 45.5  4.0
> vec2 <- 1:8                            # Another assignment operator ( <- )
> vec2                                   # Print vec2 is created with colon operator
[1] 1 2 3 4 5 6 7 8
> sqrt(vec2)                             # Find square root of vec2
[1] 1.000000 1.414214 1.732051 2.000000 2.236068 2.449490 2.645751 2.828427
> exp(1)                                 # Exponential function
[1] 2.718282
> sin(pi/2)                              # Calculates the sin x, in radians
```

Entering and executing commands in R as done in the preceding lines of code is called the Interactive Mode [3]. Via this interactive mode, we have had a rather cursory exposure to some of the concepts listed below.

- Basic arithmetic operators in R
  - + (Addition)
  - − (Subtraction)

- ^ or ** (Exponentiation)
- / (Division)

- Examples of built-in functions in R

  - **round()**
  - **min ()**
  - **max()**
  - **c()**
  - **rep()**
  - **sin()**
  - **exp()**
  - **sqrt()**

- Variables and the assignment operators (there are more than two in R, the second is the most preferred by hardcore R users)

  - **=**
  - **<-**

- Creating simple vectors via

  - **c()** (combine function)
  - **:** (colon operator)

A comprehensive discussion on these and related concepts can be explored further in one of the many excellent references highlighted in Sect. 1.1. However, the R documentation is also amazingly rich in content that knowing how to get help from it is a necessary step to proficiency in R.

1.3.2 Getting Help

There are several ways to get help in R. For a start, the manuals, resources and documentations on a myriad of functions can be accessed by typing the following in R console:

```
> help.start()
```

Hit the Enter key after typing **help.start()**, and the file pane (labelled in Fig. 1.5) will reveal links to several resources as shown in Fig. 1.6. Further, if you click on the "Home" symbol, which is pointed to using an arrow in Fig. 1.6, the file pane will also provide you with access to RStudio resources as shown in Fig. 1.7.

To know more about a specific function, a very handy method is to use a question mark followed by the name of the function. This functionality is especially useful to find more detail information about all the functions developed for this book. For

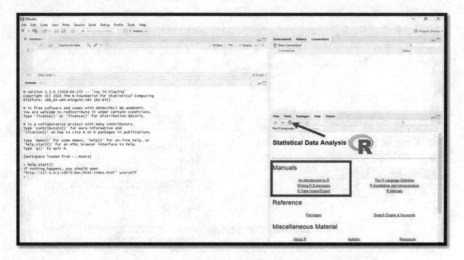

Fig. 1.6 Accessing manuals and references on R through the RStudio GUI

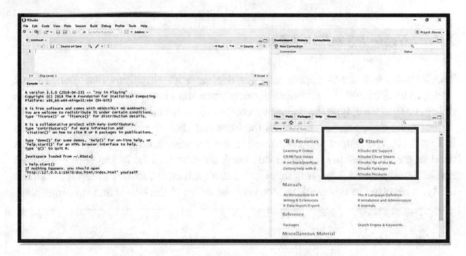

Fig. 1.7 Accessing manuals and references on RStudio through the GUI

instance, say we wish to know more about the function called plot (apparently used for plotting), we do this:

```
?plot
```

The result will be information about this function as shown in the file pane in Fig. 1.8. Meanwhile, after populating the console with many commands and you wish to clear it, use **Ctrl+L**.

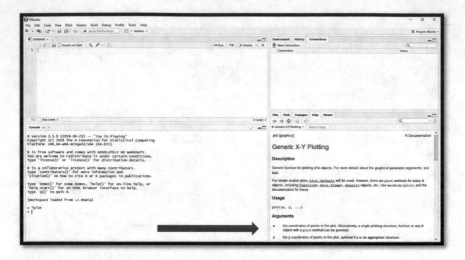

Fig. 1.8 Searching for help with "**?**"

1.3.3   Using RStudio's Editor

The RStudio's editor is appropriate for dealing with situations that demand that multiple lines of code be executed at once rather than interactively as we have been doing. For instance, type the following commands in the editor (as shown in Fig. 1.9). After typing the codes, highlight all the lines and click the "run" button as depicted in Fig. 1.10. The output will be printed in the console (see Fig. 1.11). All the codes used in the analyses presented in this book are first written in the editor and then run as demonstrated here. It is therefore a straightforward matter to replicate all the analysis reported in this book by copying the codes to the editor and running them.

```
a = 40; b = -20; d = 100                    #Define three variables
x = a + b + c                               # Assign the sum of the variables to x
vecy = c(12, 89, 21, 23, 100, 12)           # A simple vector with c()
matrixA = matrix(vecy, nrow = 2, byrow = T) # Convert vecy to a matrix
vecy[1,]                                    # Extract first element of vecy
matrixA[1,]                                 # Extract first element of matrixA
```

1.4   Packages in R

In R, packages are the "fundamental unit of shareable code" [19]. Primarily, a package is a collection of functions along with the associated documentation, but it may also include other details beyond these two.

Fig. 1.9 Multiple lines of code in RStudio Editor

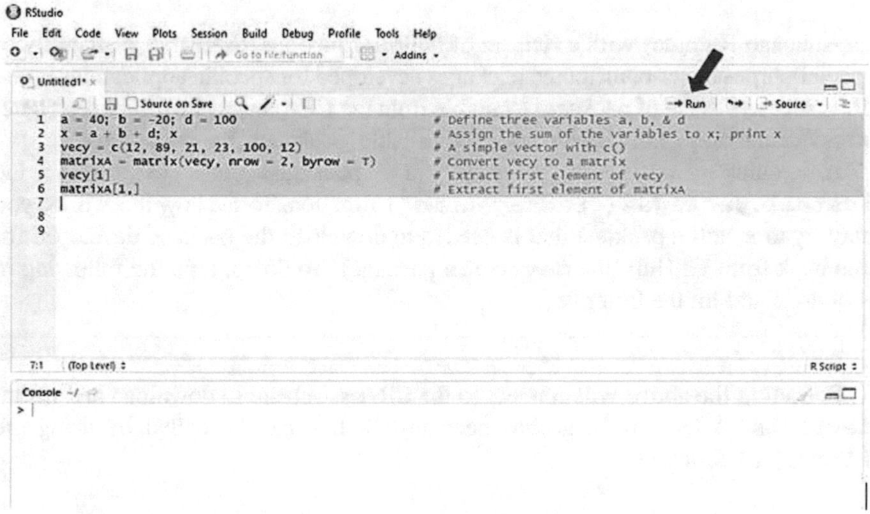

Fig. 1.10 Selecting and running multiple lines of code

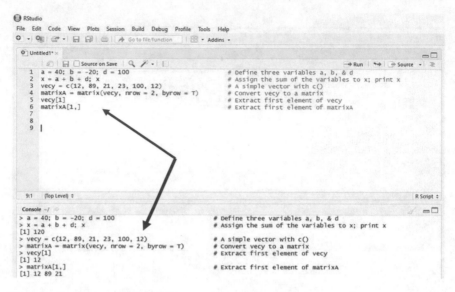

Fig. 1.11 Outputs are produced in the console for the lines of codes run in the editor

Although R comes with a rich set of built-in functions, extensive availability of special-purpose user-contributed packages developed for specific applications is one of its strength. A list of packages available from the CRAN website can be seen here: https://cran.r-project.org/web/packages/available_packages_by_name.html.

A quick way to install a package is to use the **install.packages** ("packagename") function. To see how this works, you may try to install a package that is needed to download the package developed for this book from GitHub (the **devtools** package). To do so, type the following in R console and hit the Enter key:

```
> install.packages("devtools")
```

Executing the above will connect to the CRAN website to download and install **devtools**. After a package has been installed, it can be called by using the **library()** function.

1.5   Package—MicrofiniteR

A package named **microfiniteR** is developed for this book to ease the computational tasks. As illustrated in Fig. 1.12, the package is a collection of relevant functions developed in Chaps. 2–4. There are three benefits to having the package. First, readers only need to load the package once, and all the implemented functions for the finite element models in this book will be loaded in the computer memory. Second, the description of the functions, the meanings of the input parameters and the

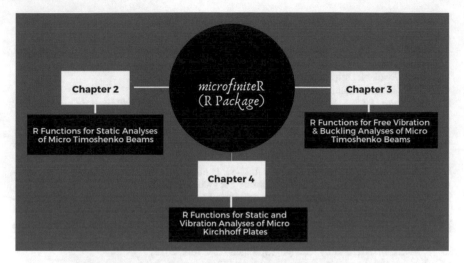

Fig. 1.12   Highlight of functions contained in `microfiniteR`

value returns by each of the developed functions can be accessed by using the query approach (question mark, followed by the function name) demonstrated in sect. 1.3.2. Third, it is likely that there will be future updates to the functions. In this case, getting the update will be a straightforward step of downloading the updated package from the code's repository on GitHub https://github.com/khameelbm/microfiniteR. Nonetheless, for first time usage, the package can be installed using the following command:

```
devtools::install_github("khameelbm/microfiniteR")
```

1.6   Summary

In this chapter, a brief historical account and a short introduction to R are provided. Instructions on how to install the required tools for R are discussed. A limited number of built-in functions in R are examined via the RStudio console and editor. A highlight of the package developed for this book is presented. A list of references that can be consulted for further explorations follows this summary.

References

1. R. Ihaka, R. Gentleman, R: a language for data analysis and graphics. J. Comput. Graph. Stat. **5,** 299–314 (1996)
2. M.J. Crawley, *The R Book* (Wiley, Hoboken, 2007)

3. N. Matloff, *The Art of R Programming: A Tour of Statistical Software Design* (No Starch Press, 2011)
4. R.C. Team, *R: A Language and Environment for Statistical Computing*, ed. Vienna, Austria: R Foundation for Statistical Computing
5. *Use R!* Available: https://www.springer.com/series/6991
6. *The R Series*. Available: https://www.crcpress.com/go/the-r-series
7. K. Soetaert, J. Cash, F. Mazzia, Differential equations, in *Solving Differential Equations in R* (Springer, 2012), pp. 1–13
8. G.W. Griffiths, *Numerical Analysis Using R* (Cambridge University Press, Cambridge, 2016)
9. V.A. Bloomfield, *Using R for Numerical Analysis in Science and Engineering* (Chapman and Hall/CRC, 2014)
10. W.E. Schiesser, *Differential Equation Analysis in Biomedical Science and Engineering: Ordinary Differential Equation Applications with R* (Wiley, Hoboken, 2014)
11. K.B. Mustapha, *Finite Element Computations in Mechanics with R: A Problem-Centered Programming Approach* (CRC Press, Boca Raton, 2018)
12. O. Jones, R. Maillardet, A. Robinson, *Introduction to Scientific Programming and Simulation Using R* (CRC Press, Boca Raton, 2009)
13. J. Albert, M. Rizzo, *R by Example* (Springer, New York, 2012)
14. R. Kabacoff, *R in Action: Data Analysis and Graphics with R* (Manning, 2015)
15. P. Teetor, *R Cookbook* (O'Reilly, 2011)
16. J. Adler, *R in a Nutshell: A Desktop Quick Reference* (O'Reilly Media, 2010)
17. H. Wickham, *Advanced R* (Taylor & Francis Limited, Abingdon, 2017)
18. M. Allerhand, *A Tiny Handbook of R* (Springer, Heidelberg, 2011)
19. H. Wickham, *R Packages* (O'Reilly Media, 2015)

Chapter 2
Bending of Microstructure-Dependent MicroBeams and Finite Element Implementations with R

Abstract The bending response of microstructure-dependent miniature beam-like structures is examined within the theoretical framework of the modified couple stress theory. Predicated on a planar assumed displacement field, and adopting the small-strain, linearly elastic and Timoshenko's shear deformable beam theory assumptions, the equilibrium equations governing the bending response of these structures are established via the variational method. Finite element solutions of the model are sought. The finite element solutions are implemented in the R programming language and then employed to illustrate the influence of size-effect on the bending response of microscale beams.

2.1 Introduction

Mechanics, like other branches of engineering science, thrives on the idealization of real physical structures with approximate physico-mechanical models.[1] Attached to these models are labels that exude the functional role of the approximation: shafts, beams, plates etc. [1]. In this chapter, the focus is on the static analysis of size-dependent structures that are approximated as microstructure-dependent beams.

Across dimensional scales (nano-, micro-scale and macro-scale), beams deliver versatile support for load-carrying or motion-generating functions, and they remain the most deployed members in mechanical systems and machines. Microstructure-dependent beams, which is of concern here, offer functional roles in microelec-tromechanical systems (MEMs). In this regard, they act in a variety of settings, from micro-scale springs (e.g. in comb-drive actuators and serpentine spring as shown in Fig. 2.1a, b), to multilayered micro-cantilevers for MEMs-based power switch (Fig. 2.1c), and as elements of resonant actuators, to probing elements in atomic force microscopy etc. [2–4]. In recent times, microbeams have also proved to be appealing in the development of micro-architected materials, where the frame comprising the unit cell of the architected material system are produced

---

[1] In the words of AEH Love in *Theoretical Mechanics,* such models facilitate *"the description of motion of material bodies".*

K. B. Mustapha, *R for Finite Element Analyses of Size-dependent Microscale Structures*, SpringerBriefs in Computational Mechanics, https://doi.org/10.1007/978-981-13-7014-4_2

13

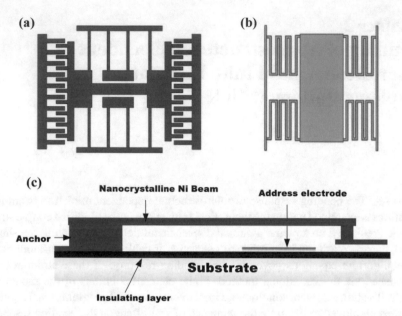

Fig. 2.1 Microscale beams in: **a** Comb-drive actuator; **b** serpentine spring; **c** MEMs-based power switch

with microscale geometry (Fig. 2.2) [5–7]. In all the aforementioned applications, knowledge of the static capability of microscale beams in flexural loading becomes crucial for effective deployment.

In Sect. 2.2, the distributed mathematical model of microscale beams with a size-dependent feature is studied based on the modified couple stress theory. Next, the finite element formulation of the presented model follows in Sect. 2.3. A collection of R functions implemented to support the numerical evaluation of the model resides in Sect. 2.4, and this is followed by a summary of the chapter and some selected references.

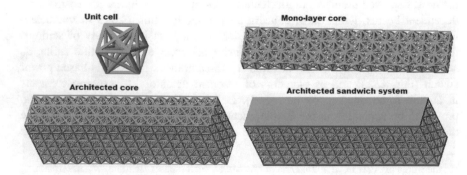

Fig. 2.2 Microscale beams as members of the unit cells in micro-architected materials

2.2 Governing Differential Equations

The mathematical models of structures treated in this text are obtained via the variational method. Procedurally, this approach hinges on the knowledge of a few fundamental concepts: the displacement-strain relations, the stress-strain relations, the strain energy functional and the statement of the total potential energy [8]. Further, each of these relations come with some notable assumptions to render the problem tractable. Accordingly, in what follows, we shall first start with the form of the strain energy density function through which the small-scale effect[2] of microscale structures is considered. We will then move on to the kinematic relations, the constitutive rule, and finally the establishment of the linear equations of motion via the principle of minimum potential energy.

2.2.1 Classical Versus Non-classical Strain Density Function

In the classical theory of finite elasticity, the strain energy density[3] (W) is taken to be a function of strain (ε):

$$W = W(\varepsilon) \tag{2.1}$$

It turns out the definition of the strain energy density in Eq. (2.1) is insufficient to account for observed phenomena that happens in certain classes of materials [9]. For instance, the mathematical models derived from it are incapable of predicting size-dependent phenomenological traits such as dispersion strengthening and the Hall-petch effect exhibited by some structures. Examples of such structures include microscale bodies for which the geometric dimensions approach the internal material length scale [10], microstructured porous materials and ductile cellular solids [11–14]. Indeed, experimental evidences for size-specific behaviour in micro-scale solids have been observed and discussed in several studies, notable among which are the investigations by Kelly [15], Ebeling and Ashby [16], Ashby [17], Hutchinson [18] and Lam and Chong [19].

An important question that arises from the aforementioned limitation is how to incorporate the microstructural effect in the modelling of microstructure-dependent bodies. The quests for a solution led to the emergence of generalized elasticity theories. Efforts in this direction date back to many decades as documented in the historical accounts of the subject matter by Askes and Aifantis [20], Dyszlewicz [21], Thai, et al. [22] and Khakalo and Niiranen [23]. Nonetheless, a partial list of contributions includes the work of Voigt [24], the Cosserat

---

[2]The words size-effect and small-scale effect are used interchangeably in this text. In the context of this book, they are used to describe the dependency of the response of small-scale structures on their size.

[3]A scalar quantity typically measured in energy per volume.

brothers [25], Mindlin [26], Toupin [27], Koiter [28], Eringen [29], and Eringen and Suhubi [30].

In a series of expositions, Mindlin proposed the linear versions of the generalized strain energy density function, forming the basis of what is now widely known as the strain gradient theory [31–33]. Models derived from the strain gradient theories have been employed for the description of the mechanical response of many micro- and nano-scale structures [34, 35]. The non-linear version of this theory appeared first in Toupin [27]. In one of the formulations provided by Mindlin, the strain energy density is redefined so that it is mapped to the strain tensor, as well as its first and second gradients. That is:

$$W = W(\boldsymbol{\varepsilon}, \nabla \boldsymbol{\varepsilon}, \nabla \nabla \boldsymbol{\varepsilon}) \tag{2.2}$$

However, Eq. (2.2) turns out to be dauntingly characterized by 54 independent variables and 5 material parameters. A variant of Eq. (2.2) considers the strain energy density function to be only a function of the strain and the second gradient of the displacement, leading to:

$$W = W(\varepsilon_{ij}, \kappa_{ijk}) \tag{2.3}$$

where ε_{ij} and κ_{ijk} symbolize the components of the strain tensor and the second gradient of the displacement, respectively. Written in full form, Eq. (2.3) becomes:

$$W = \frac{1}{2}\lambda\varepsilon_{ii}\varepsilon_{jj} + \mu\varepsilon_{ij}\varepsilon_{ij} + a_1\kappa_{iik}\kappa_{kjj} + a_2\kappa_{ijj}\kappa_{ikk} + a_3\kappa_{iik}\kappa_{jjk}$$
$$+ a_4\kappa_{ijk}\kappa_{ijk} + a_5\kappa_{ijk}\kappa_{kji} \tag{2.4}$$

where the parameters $a_p(p = 1, 2, \ldots 5)$ denote five additional elastic constants associated with the invariants of the second deformation gradients. Determining these additional parameters remains an open problem. Accordingly, there have been a few proposed modified versions of the Mindlin's strain gradient theory, one of which is [36, 37]:

$$W_m = \frac{1}{2}\lambda\varepsilon_{ii}\varepsilon_{jj} + \mu\varepsilon_{ij}\varepsilon_{ij} + l_0\varepsilon_{mm,i}\varepsilon_{nn,j} + l_1\gamma_{ijk}^{(1)}\gamma_{ijk}^{(1)} + l_2\chi_{ij}^s\chi_{ij}^s \tag{2.5}$$

where l_0, l_1 and l_2 are related to the dilatation gradients, deviatoric gradients and the symmetric parts of rotation gradients, respectively. If one is to simplify things further for practical purposes by assuming $l_0 = l_1 = 0$, then a one-parameter theory called the modified couple stress theory (MCST) emerges [36, 37]. MCST has gained traction in recent years among researchers with interest in the mechanics of small-scale structures. Key to its popularity is the perceived mathematical and experimental convenience it offers in providing a way to quantify the modified stiffness for a wide variety of small-scale structures under different loading regimes.

In accordance with the MCST, the size-dependent strain energy (Π_s) stored in a deformed isotropic linear elastic body manifold occupying a volume Ω is characterized by two parts. One part is associated with the classical stress components and the other due to the couple stress components [37, 38]:

$$\Pi_s = \frac{1}{2} \int_\Omega \left(\sigma_{ij} \varepsilon_{ij} + m_{ij} \chi_{ij} \right) d\Omega \qquad (2.6)$$

where σ_{ij} denote the components of the classical stress tensors. Accordingly, m_{ij} and χ_{ij} refer to the components of the couple stress and the symmetric curvature tensors, respectively.

2.2.2 Displacement-Strain Relations

Expressed through the derivatives of components of the displacement vector (u_i), the components of the strain tensor are obtained from the displacement-strain relation as:

$$\varepsilon_{ij} = \frac{1}{2} \left(\frac{\partial u_i}{\partial x_j} + \frac{\partial u_j}{\partial x_i} \right) \qquad (2.7)$$

Likewise, using the components of the rotation vector (θ_i), the components of the symmetric curvature tensor are obtained from:

$$\chi_{ij} = \frac{1}{2} (\theta_{i,j} + \theta_{j,i}) \quad \text{where } \theta_i = \frac{1}{2} e_{ijk} u_{k,j} \qquad (2.8)$$

2.2.3 Stress-Strain Relations

The classical and the couple stress fields (σ_{ij} and m_{ij}) are mapped to the displacement field through the following constitutive rules:

$$\sigma_{ij} = \lambda \, \varepsilon_{mm} \delta_{ij} + 2\mu \varepsilon_{ij} \qquad (2.9)$$

$$m_{ij} = 2l^2 \mu \chi_{ij} \qquad (2.10)$$

where, as Eq. (2.9) reveals, the stress field is mapped to the deformation field through the usual material parameters λ and μ, which are the bulk and shear modulus, respectively. In a similar spirit, the higher-order couple stress field (Eq. 2.10) is mapped to the deformation field through an additional material constant in the form of l. This new parameter, as enunciated in Ma et al. [39], can be determined from torsion tests of thin cylinders of varying diameters or bending tests of microscale beam of different thickness as described in Chong et al. [40]. In all, Eqs. (2.6) through (2.10) facilitate the establishment of the governing equations for microstructure-dependent structures based on the MCST.

2.2.4 Timoshenko Beam Theory and Reduced Displacement Field

All load-carrying structural elements are three-dimensional entities. However, via the concept of assumed displacement or stress fields, representations of many structures can be simplified by using a two-dimensional or a one-dimensional equivalent[4] [41, 42].

Let us consider a prismatic microscale beam of length L, for which we wish to obtain the linear governing equations. The beam has a rectangular constant cross-section with geometric dimensions b and h. Adopting the Timoshenko beam theory, one takes the reduced displacement field to be [43]:

$$u_1(x, y, z) = -z\phi(x); u_2(x, y, z) = 0; u_3(x, y, z) = w(x) \qquad (2.11)$$

where u_1 is the axial displacement of the beam induced by bending, while the lateral displacements along the y-axis and z-axis are denoted by u_2 and u_3, respectively. The cross-sectional rotation (with respect to a transverse normal line) is denoted by ϕ, and it is taken to be uncoupled from the slope of the deformed shape of the beam. Further, w is the total transverse displacement due to shear and bending deformations. Technically, Eq. (2.11) is contingent on two kinematic assumptions. First, the cross-section of the beam is taken to be rigid on its own plane, and this guarantees the adoption of a planar deformation of the beam, say under the effect of a distributed load $q(x)$. Second, the cross-section rotates around a neutral axis, remaining plane but non-orthogonal to it.

By using Eq. (2.11) with respect to Eq. (2.7), one obtains the following non-zero strain components[5]:

$$\varepsilon_{xx} = \frac{du_1}{dx} = -z\frac{d\phi}{dx}; \qquad (2.12)$$

$$\varepsilon_{xz} = \varepsilon_{zx} = \frac{\gamma_{xz}}{2} = \frac{1}{2}\left(\frac{du_1}{dz} + \frac{du_3}{dx}\right) = \frac{1}{2}\left(-\phi + \frac{dw}{dx}\right) \qquad (2.13)$$

---

[4]In two-dimensional approximations, the field variable (such as displacement or stress field) depends on two coordinates, and the boundary conditions are imposed on a line. Conversely, in one-dimensional approximations, the field variable is a function of one coordinate and the boundary conditions are imposed on points.

[5]The displacement-strain relations are obtained with the assumption that the axial, shear and curvature strains along with the rotation of the beam are small in comparison to unity.

Furthermore, evaluating Eqs. (2.8)–(2.10) with the reduced displacement field of Eq. (2.11), one retrieves the following non-trivial components of the curvature, stress and couple stress tensors [39, 44]:

$$\chi_{xy} = \chi_{yx} = \frac{1}{4}\left(-\frac{d\phi}{dx} - \frac{d^2w}{dx^2}\right) \tag{2.14}$$

$$\sigma_{xx} = E\varepsilon_{xx} = -Ez\frac{d\phi}{dx}; \quad \sigma_{xz} = 2G\varepsilon_{xz} = \mu\left(-\phi + \frac{dw}{dx}\right) \tag{2.15}$$

$$m_{xy} = m_{yx} = 2Gl^2\chi_{xy} = \frac{Gl^2}{2}\left(-\frac{d\phi}{dx} - \frac{d^2w}{dx^2}\right) \tag{2.16}$$

where E and G represent the Young's and the shear modulus,[6] respectively. Plugging of the components in Eqs. (2.12)–(2.16) into Eq. (2.6) provides the strain energy of the microscale Timoshenko beam as:

$$
\begin{aligned}
\Pi_s &= \frac{1}{2}\int_0^L\int_A (\sigma_{xx}\varepsilon_{xx} + 2\sigma_{xz}\varepsilon_{xz})\,dA\,dx + \frac{1}{2}\int_0^L\int_\Omega 2m_{xy}\chi_{xy}\,dA\,dx \\
&= \frac{1}{2}\int_0^L\int_A\left[E\left(z\frac{d\phi}{dx}\right)^2 + G\left(-\phi + \frac{dw}{dx}\right)^2\right]dA\,dx \\
&\quad + \frac{1}{2}\int_0^L\int_A\left[\frac{Gl^2}{4}\left(-\frac{d\phi}{dx} - \frac{d^2w}{dx^2}\right)^2\right]dA\,dx
\end{aligned}
\tag{2.17}
$$

By introducing the shear correction factor[7] (k), $\int_A z^2\,dA = I$; $\int_A dA = A$, then Eq. (2.17) becomes:

$$
\begin{aligned}
\Pi_s &= \frac{1}{2}\int_0^L EI\left(\frac{d\phi}{dx}\right)^2 dx + \frac{1}{2}\int_0^L GAk\left(-\phi + \frac{dw}{dx}\right)^2 dx \\
&\quad + \frac{1}{2}\int_0^L \frac{Gl^2A}{4}\left(-\frac{d\phi}{dx} - \frac{d^2w}{dx^2}\right)^2 dx
\end{aligned}
\tag{2.18}
$$

The total work done[8] by the distributed load $q(x)$ acting at the centroid of the beam between the end points $[0, L]$ is:

---

[6]We have replaced μ, originally in Eq. (2.10), by G.

[7]This is to remedy the fact that the shear stress has been taken to be independent of z, and hence uniform across the thickness.

[8]The total work done is also a scalar quantity like the strain energy.

$$\Pi_W = - \int_0^L q(x)w\,dx \tag{2.19}$$

Combining Eqs. (2.18) and (2.19) gives the total potential energy for microscale Timoshenko beams with isotropic material properties as:

$$\begin{aligned} \Pi &= \Pi_s + \Pi_W \\ &= \frac{1}{2} \int_0^L \left[EI\left(\frac{d\phi}{dx}\right)^2 + GAk\left(-\phi + \frac{dw}{dx}\right)^2 \right. \\ &\quad \left. + \frac{Gl^2 A}{4}\left(-\frac{d\phi}{dx} - \frac{d^2w}{dx^2}\right)^2 - q(x)w(x) \right] dx \end{aligned} \tag{2.20}$$

In a compact form, we can re-write Eq. (2.20) in the form of a functional F, such that:

$$\Pi = \int_0^L F\big(x, \phi, w, \phi', w', w''\big)\,dx \tag{2.21}$$

where the prime (′) represents a derivative with respect to x. It is now possible to obtain the governing equations of the structure by invoking the rules of calculus of variations, which stipulates that the extremum of Π shall be obtained if the following two Euler-Lagrange differential equations are satisfied:

$$\frac{\partial F}{\partial \phi} - \frac{d}{dx}\frac{\partial F}{\partial \phi'} = 0 \tag{2.22}$$

$$\frac{\partial F}{\partial w} - \frac{d}{dx}\frac{\partial F}{\partial w'} + \frac{d^2}{dx^2}\frac{\partial F}{\partial w''} = 0 \tag{2.23}$$

where

$$\begin{aligned} \frac{\partial F}{\partial \phi} &= GAk\big(-\phi + w'\big); \quad \frac{\partial F}{\partial \phi'} = -EI\,\phi' - \frac{1}{4}GAl^2\big(\phi' + w''\big) \\ \frac{\partial F}{\partial w} &= -q; \quad \frac{\partial F}{\partial w'} = GAk\big(\phi - w'\big); \quad \frac{\partial F}{\partial w''} = -\frac{1}{4}GAl^2\big(\phi' + w''\big) \end{aligned} \tag{2.24}$$

Plugging of Eqs. (2.24) into Eqs. (2.22) and (2.23) leads to the governing equations as:

$$GAk\left(-\phi + \frac{dw}{dx}\right) + EI\frac{d^2\phi}{dx^2} + \frac{GAl^2}{4}\left(\frac{d^2\phi}{dx^2} + \frac{d^3w}{dx^3}\right) = 0 \tag{2.25}$$

$$GAk\left(-\frac{d\phi}{dx}+\frac{d^2w}{dx^2}\right)-\frac{GAl^2}{4}\left(\frac{d^3\phi}{dx^3}+\frac{d^4w}{dx^4}\right)=q(x) \qquad (2.26)$$

where the equations hold on $x \in \le (0, L)$. The first variation of Eq. (2.20) yields the boundary conditions to be imposed at the ends of the beam, and it demands that at $x = 0$ and $x = L$:

- Either ϕ is specified or $-EI\frac{d\phi}{dx}-\frac{GAl^2}{4}\left(\frac{d\phi}{dx}+\frac{d^2w}{dx^2}\right)=0$ \qquad (2.27)

- Either w is specified or $GAk\left(\phi-\frac{dw}{dx}\right)+\frac{GAl^2}{4}\left(\frac{d^2\phi}{dx^2}+\frac{d^3w}{dx^3}\right)=0$

$$(2.28)$$

- Either $\frac{dw}{dx}$ is specified or $-\frac{GAl^2}{4}\left(\frac{d\phi}{dx}+\frac{d^2w}{dx^2}\right)=0$ \qquad (2.29)

2.3 Finite Element Formulations

The finite element method (FEM) has dominated the process of design and analysis in the engineering field for the last 7 decades. Within this time lapse, many variants of the methods have emerged and lead to the grouping of the formulation based on two criteria. The first categorization relates to the nature of the control imposed on the parameters of the finite-dimensional space (mesh size and the order of the approximating polynomial). In this regards, there is: (i) the p-version of the FEM, where concern is on finding the accurate solution of a function by fixing the mesh, while increasing the order of the approximation function [45]; (ii) the h-version of the FEM, where accuracy of the solution depends on fixing the order of the polynomial, while decreasing the size of the mesh; and (iii) the hp-version where the solution accuracy relies on the concurrent varying of the mesh and the approximating polynomials [45, 46].

The formulation in this book follows the h-version of the FEM. We will consider two models of the Timoshenko beam element established on the basis of the modified couple stress theory. The first model, a 6 degree of freedom model, considers at the node the transverse displacement (w), the first derivative of w and the cross-sectional rotation (ϕ) as the nodal degrees of freedom [47]. In the second model, a 4 degree of freedom model, we consider only the transverse displacement (w) and the cross-sectional rotation (ϕ) in consonant with Kahrobaiyan et al. [44].

2.3.1 The 6 Degrees of Freedom Finite Element Model

The derivatives of the transverse displacement and the cross-section rotation in the governing equations have 4 and 3, respectively, as the highest order. This automatically requires that a two-node beam element based on these established governing equations shall have, as its nodal degree of freedom (d.o.f), transverse displacement (w), slope of the displacement $\left(\frac{dw}{dx}\right)$, and rotation ($\phi$). With this argument, a 6 degree of freedom two-node beam element thus arises. According to Arbind and Reddy [48] and Dehrouyeh-Semnani and Bahrami [47], the 6 d.o.f model is obtained by specifying independent interpolation functions for w and ϕ. In this spirit, a cubic and a linear distribution is assumed for the displacement w and ϕ over an element as:

$$w = c_1 + c_2 x + c_3 x^2 + c_4 x^3 \tag{2.30}$$

$$\psi = \frac{dw}{dx} = c_2 + 2c_3 x + 3c_4 x^2 \tag{2.31}$$

$$\phi = c_5 + c_6 x \tag{2.32}$$

where $c_1 - c_6$ are unknown constants (the generalized coordinates) that will be determined by imposing the following boundary conditions on Eqs. (2.30–2.32):

- At node 1, $x = 0$: $w = w_1$; $\psi = \psi_1$; $\phi = \phi_1$
- At node 2, $x = L$: $w = w_2$; $\psi = \psi_2$; $\phi = \phi_2$

Evaluating of which leads to a system of equations summarized in the matrix form as:

$$\begin{Bmatrix} w_1 \\ \psi_1 \\ \phi_1 \\ w_2 \\ \psi_2 \\ \phi_2 \end{Bmatrix} = \begin{bmatrix} 1 & 0 & 0 & 0 & 0 & 0 \\ 0 & 1 & 0 & 0 & 0 & 0 \\ 0 & 0 & 0 & 0 & 1 & 0 \\ 1 & L & L^2 & L^3 & 0 & 0 \\ 0 & 1 & 2L & 3L^2 & 0 & 0 \\ 0 & 0 & 0 & 0 & 1 & L \end{bmatrix} \begin{Bmatrix} c_1 \\ c_2 \\ c_3 \\ c_4 \\ c_5 \\ c_6 \end{Bmatrix} \tag{2.33}$$

In view of Eq. (2.33), the vector of the unknown coefficients $\{c\}$ are determined as:

$$\{c\} = [M_6]^{-1}\{w_e\}_6 \tag{2.34}$$

where

$$[M_6]^{-1} = \begin{bmatrix} 1 & 0 & 0 & 0 & 0 & 0 \\ 0 & 1 & 0 & 0 & 0 & 0 \\ -\frac{3}{L^2} & -\frac{2}{L} & 0 & \frac{3}{L^2} & -\frac{1}{L} & 0 \\ \frac{2}{L^3} & \frac{1}{L^2} & 0 & -\frac{2}{L^3} & \frac{1}{L^2} & 0 \\ 0 & 0 & 1 & 0 & 0 & 0 \\ 0 & 0 & -\frac{1}{L} & 0 & 0 & \frac{1}{L} \end{bmatrix}; \{w_e\}_6 = \begin{Bmatrix} w_1 \\ \psi_1 \\ \phi_1 \\ w_2 \\ \psi_2 \\ \phi_2 \end{Bmatrix} \tag{2.35}$$

With the aid of Eq. (2.34), we re-write Eqs. (2.30–2.32) as:

$$w = \begin{bmatrix} 1 & x & x^2 & x^3 & 0 & 0 \end{bmatrix}\{c_i\}$$
$$= \begin{bmatrix} 1 & x & x^2 & x^3 & 0 & 0 \end{bmatrix} [M_6]^{-1}\{w_e\}_6 \tag{2.36}$$

$$\psi = \begin{bmatrix} 0 & 1 & 2x & 3x^2 & 0 & 0 \end{bmatrix}\{c_i\}$$
$$= \begin{bmatrix} 0 & 1 & 2x & 3x^2 & 0 & 0 \end{bmatrix} [M_6]^{-1}\{w_e\}_6 \tag{2.37}$$

$$\phi = \begin{bmatrix} 0 & 0 & 0 & 0 & 1 & x \end{bmatrix}\{c_i\} = \begin{bmatrix} 0 & 0 & 0 & 0 & 1 & x \end{bmatrix} [M_6]^{-1}\{w_e\}_6 \tag{2.38}$$

Evaluating the underlined terms leads to:

$$w = \underbrace{\left[1 - \frac{3x^2}{L^2} + \frac{2x^3}{L^3} \quad \left(x - \frac{2x^2}{L} + \frac{x^3}{L^2}\right) \quad 0 \quad \left(\frac{3x^2}{L^2} - \frac{2x^3}{L^3}\right) \quad \left(-\frac{x^2}{L} + \frac{x^3}{L^2}\right) \quad 0 \right]}_{[N_w]}\{w_e\}_6$$
$$\tag{2.39}$$

$$\psi = \underbrace{\left[-\frac{6x}{L^2} + \frac{6x^2}{L^3} \quad 1 - \frac{4x}{L} + \frac{3x^2}{L^2} \quad 0 \quad \frac{6x}{L^2} - \frac{6x^2}{L^3} \quad -\frac{2x}{L} + \frac{3x^2}{L^2} \quad 0 \right]}_{[N_\psi]}\{w_e\}_6 \tag{2.40}$$

$$\phi = \underbrace{\left[0 \quad 0 \quad 1 - \frac{x}{L} \quad 0 \quad 0 \quad \frac{x}{L} \right]}_{[N_\phi]}\{w_e\}_6 \tag{2.41}$$

where $\{w_e\}_6$ is a vector of the nodal d.o.f. Further, $[N_w]$, $[N_\psi]$ and $[N_\phi]$ denote the vector of interpolation functions for the transverse displacement, the slope and the cross-section rotation, respectively.

The non-trivial components of the strain and rotation gradient tensor can be written in terms of the following B matrices:

$$[B_\phi] = \frac{d[N_\phi]}{dx} = \begin{bmatrix} 0 & 0 & -\frac{1}{L} & 0 & 0 & \frac{1}{L} \end{bmatrix} \tag{2.42}$$

$$[B_\gamma] = \left(-[N_\phi] + \frac{d[N_w]}{dx} \right)$$
$$= \begin{bmatrix} \frac{6x(-L+x)}{L^3} & \frac{L^2 - 4Lx + 3x^2}{L^2} \end{bmatrix}$$

$$-1 + \frac{x}{L} \quad \frac{6(L-x)x}{L^3} \quad \frac{x(-2L+3x)}{L^2} \quad -\frac{x}{L} \;] \tag{2.43}$$

$$[B_\chi] = \frac{1}{2}\left(-\frac{d[N_\phi]}{dx} - \frac{d^2[N_w]}{dx^2}\right)$$

$$= [\; \frac{3(L-2x)}{L^3} \quad \frac{2L-3x}{L^2} \quad \frac{1}{2L}$$

$$-\frac{3(L-2x)}{L^3} \quad \frac{L-3x}{L^2} \quad -\frac{1}{2L} \;] \tag{2.44}$$

Therefore:

$$\{\varepsilon_{xx}\} = -z[B_\phi]_6\{w_e\}_6; \{\varepsilon_{xz}\}$$

$$= [B_\gamma]_6\{w_e\}_6; \{\chi_{xy}\}$$

$$= \frac{1}{2}[B_\chi]_6\{w_e\}_6 \tag{2.45}$$

$$\{\sigma_{xx}\} = -Ez[B_\phi]_6\{w_e\}_6; \{\sigma_{xz}\}$$

$$= Gk[B_\gamma]_6\{w_e\}_6; \{m_{xy}\}$$

$$= Gl^2[B_\chi]_6\{w_e\}_6 \tag{2.46}$$

Bearing in mind Eqs. (2.45)–(2.46), we re-formulate the strain energy (Π_s) and the work done by external forces (W) on the element as:

$$\Pi_s = \frac{1}{2}\int_0^{L_e}\left[EI\{\sigma_{xx}\}^T\{\varepsilon_{xx}\} + GAk\{\sigma_{xz}\}^T\{\varepsilon_{xz}\} + Gl^2A\{m_{xy}\}^T\{\chi_{xy}\}\right]dx$$

$$= \frac{EI}{2}\int_0^{L_e}([B_\phi]\{w_e\}_6)^T([B_\phi]\{w_e\}_6)\,dx$$

$$+ \frac{kGA}{2}\int_0^{L_e}([B_\gamma]\{w_e\}_6)^T([B_\gamma]\{w_e\}_6)\,dx$$

$$+ \frac{Gl^2A}{2}\int_0^{L_e}([B_\chi]\{w_e\}_6)^T([B_\chi]\{w_e\}_6])\,dx \tag{2.47}$$

$$\Pi_{exW} = \{w_e\}_6^T\{f_e\} \tag{2.48}$$

where Π_{exW} is the potential energy of the applied load. The elastostatic stiffness matrix of the 6 d.o.f microscale Timoshenko beam is then obtained by minimizing the total potential energy Π with respect to $\{w_e\}_6$ as:

$$\frac{\partial \Pi}{\partial \{w_e\}^T} = 0$$

$$= \left(EI \int_0^{L_e} [B_\phi]^T [B_\phi] \, dx + kGA \int_0^{L_e} [B_\gamma]^T [B_\gamma] \, dx \right.$$

$$\left. + Gl^2 A \int_0^{L_e} [B_\chi]^T [B_\chi] \, dx \right) \{w_e\}_6 - \{f_e\} \tag{2.49}$$

From which one obtains:

$$\left(EI \int_0^{L_e} [B_\phi]^T [B_\phi] \, dx + kGA \int_0^{L_e} [B_\gamma]^T [B_\gamma] \, dx \right.$$

$$\left. + Gl^2 A \int_0^{L_e} [B_\chi]^T [B_\chi] \, dx \right) \{w_e\}_6 = \{f_e\} \tag{2.50}$$

In full form, the finite element matrix equation for the 6-dof Timoshenko beam element then takes the form:

$$\{f_e\}_6 = [K_{6DOF}]\{w_e\}_6 \tag{2.51}$$

where, if we let $kAG = \alpha_1$, $GAl^2 = \alpha_2$ and $EI = \alpha_3$, then:

$[K_{6DOF}]$

$$= \begin{bmatrix}
\frac{6\alpha_1}{5L} + \frac{3\alpha_2}{L^3} & \frac{\alpha_1}{10} + \frac{3\alpha_2}{2L^2} & \frac{\alpha_1}{2} & -\frac{6\alpha_1}{5L} - \frac{3\alpha_2}{L^3} & \frac{\alpha_1}{10} + \frac{3\alpha_2}{2L^2} & \frac{\alpha_1}{2} \\
\frac{\alpha_1}{10} + \frac{3\alpha_2}{2L^2} & \frac{2L\alpha_1}{15} + \frac{\alpha_2}{L} & -\frac{L\alpha_1}{12} + \frac{\alpha_2}{4L} & -\frac{\alpha_1}{10} - \frac{3\alpha_2}{2L^2} & -\frac{L\alpha_1}{30} + \frac{\alpha_2}{2L} & \frac{L\alpha_1}{12} - \frac{\alpha_2}{L} \\
\frac{\alpha_1}{2} & -\frac{L\alpha_1}{12} + \frac{\alpha_2}{4L} & \left(\frac{L\alpha_1}{3} + \frac{\alpha_2}{4L} + \frac{\alpha_3}{L}\right) & -\frac{\alpha_1}{2} & \frac{L\alpha_1}{12} - \frac{\alpha_2}{4L} & \frac{L\alpha_1}{6} - \frac{\alpha_2}{4L} - \frac{\alpha_3}{L} \\
-\frac{6\alpha_1}{5L} - \frac{3\alpha_2}{L^3} & -\frac{\alpha_1}{10} - \frac{3\alpha_2}{2L^2} & -\frac{\alpha_1}{2} & \frac{6\alpha_1}{5L} + \frac{3\alpha_2}{L^3} & -\frac{\alpha_1}{10} - \frac{3\alpha_2}{2L^2} & -\frac{\alpha_1}{2} \\
\frac{\alpha_1}{10} + \frac{3\alpha_2}{2L^2} & -\frac{L\alpha_1}{30} + \frac{\alpha_2}{2L} & \frac{L\alpha_1}{12} - \frac{\alpha_2}{4L} & -\frac{\alpha_1}{10} - \frac{3\alpha_2}{2L^2} & \frac{2L\alpha_1}{15} + \frac{\alpha_2}{L} & -\frac{L\alpha_1}{12} + \frac{\alpha_2}{4L} \\
\frac{\alpha_1}{2} & \frac{L\alpha_1}{11} - \frac{\alpha_2}{4L} & \frac{L\alpha_1}{6} - \frac{\alpha_2}{4L} - \frac{\alpha_3}{L} & -\frac{\alpha_1}{2} & -\frac{L\alpha_1}{12} + \frac{\alpha_2}{4L} & \left(\frac{L\alpha_1}{3} + \frac{\alpha_2}{4L} + \frac{\alpha_3}{L}\right)
\end{bmatrix}$$

$$\tag{2.52}$$

2.3.2 The 4 Degrees of Freedom Reduced-Order Model

In what follows, we shall present a reduced-order model of the two-node Timoshenko beam element based on the modified couple stress theory. The reduced-order model has 2 d.o.f at each node as shown in Fig. 2.3.

To understand why the reduced-order model may be appropriate, we will cast the governing equations established earlier into a non-dimensional form through the use of the following:

$$q(x) = 0; \xi = x/L; g = \frac{EI}{GAkL^2}; \beta^2 = \frac{l^2}{4kL^2} \tag{2.53}$$

By substituting (2.53) into Eqs. (2.25–2.26), one obtains:

$$\left(-\phi + \frac{dw}{d\xi}\right) + g\frac{d^2\phi}{d\xi^2} + \beta^2\left(\frac{d^2\phi}{d\xi^2} + \frac{d^3w}{d\xi^3}\right) = 0 \tag{2.54}$$

$$\left(-\frac{d\phi}{d\xi} + \frac{d^2w}{d\xi^2}\right) - \beta^2\left(\frac{d^3\phi}{d\xi^3} + \frac{d^4w}{d\xi^4}\right) = 0 \tag{2.55}$$

where the material length scale parameter (l), typically a few microns for most materials as elaborated in Kahrobaiyan et al. [44], is now couched in parameter β^2. Indeed, a widely used value for l in the case of epoxy polymer is 17.6 μm [36]. Further, it is revealed in the recent experimental studies by Liebold and Müller [49] and Li et al. [50], that values smaller than 17.6 μm are possible for materials such as SU-8 polymer, titanium and copper. Premised on the smallness of the ratio $\left(\frac{l}{L}\right)^2$, and hence β^2, the underlined terms in (2.54–2.55) can be ignored to obtain a set of reduced-order equations:

$$\left(-\phi + \frac{dw}{d\xi}\right) + g\frac{d^2\phi}{d\xi^2} = 0 \tag{2.56}$$

$$\left(-\frac{d\phi}{d\xi} + \frac{d^2w}{d\xi^2}\right) = 0 \tag{2.57}$$

Obviously, the reduced-order equations allow the transverse displacement (w) to be interpolated by a cubic polynomial function as done earlier:

$$w = a_1 + a_2x + a_3x^2 + a_4x^3 \tag{2.58}$$

Now, from Eq. (2.56), one recognizes the first term as the shear strain, $\gamma = \left(-\phi + \frac{dw}{d\xi}\right)$. In consonant with the assumption of a constant shear strain across the

Fig. 2.3 The reduced-order two-node Timoshenko microscale beam element

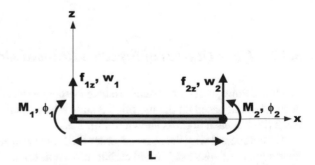

cross-section of the microscale beam, we may allow γ to be a generic constant c [51]. Thus, Eq. (2.56) becomes:

$$c + g\frac{d^2\phi}{d\xi^2} = 0 \qquad (2.59)$$

From (2.57), we have:

$$\frac{d\phi}{d\xi} = \frac{d^2w}{d\xi^2} \qquad (2.60)$$

Combining 2.59 and 2.60, bearing in mind (2.58), leads to:

$$c = -g\frac{d^2\phi}{d\xi^2} = -g\frac{d^3w}{d\xi^3} = -6ga_4 \qquad (2.61)$$

The interpolation for ϕ is then derived as[9]:

$$\phi = \frac{dw}{d\xi} + c = \frac{dw}{d\xi} - 6a_4g \qquad (2.62)$$

By obtaining the first derivative of w and substituting in 2.62, one has ϕ in terms of constants a_2, a_3 and a_4:

$$\phi = a_2 + 2a_3x + \left(3x^2 + 6g\right)a_4 \qquad (2.63)$$

where $a_1 - a_4$ are unknown constants to be found, again, by imposing the following boundary conditions on Eqs. (2.58 and 2.63):

- At node 1, $x = 0$: $w = w_1$; $\phi = \phi_1$
- At node 2, $x = L$: $w = w_2$; $\phi = \phi_2$

It is to be noted that Eqs. (2.58) and (2.63) have both been written in terms of x rather than ξ to facilitate dealing with the dimensional form of the governing equations.

By evaluating with the stated boundary conditions, a system of equations emerged that can be summarized as:

$$\begin{Bmatrix} w_1 \\ \phi_1 \\ w_2 \\ \phi_2 \end{Bmatrix} = \underbrace{\begin{bmatrix} 1 & 0 & 0 & 0 \\ 0 & 1 & 0 & 6g \\ 1 & L & L^2 & L^3 \\ 0 & 1 & 2L & 6g+3L^2 \end{bmatrix}}_{[M_4]} \begin{Bmatrix} a_1 \\ a_2 \\ a_3 \\ a_4 \end{Bmatrix} \qquad (2.64)$$

---

[9] From the knowledge that $\gamma = \left(-\phi + \frac{dw}{d\xi}\right) = c$.

From which, the vector of unknown coefficients $\{a\}$ are determined as:

$$\{a\} = [M_4]^{-1}\{w_e\}_4 \tag{2.65}$$

where

$$[M_4]^{-1} = \begin{bmatrix} 1 & 0 & 0 & 0 \\ -\frac{12gL}{12gL^2+L^4} & \frac{6gL^2+L^4}{12gL^2+L^4} & \frac{12gL}{12gL^2+L^4} & -\frac{6gL^2}{12gL^2+L^4} \\ -\frac{3gL^2}{12gL^2+L^4} & \frac{-6gL-2L^3}{12gL^2+L^4} & \frac{3L^2}{12gL^2+L^4} & \frac{6gL-L^3}{12gL^2+L^4} \\ \frac{2L}{12gL^2+L^4} & \frac{L^2}{12gL^2+L^4} & -\frac{2L}{12gL^2+L^4} & \frac{L^2}{12gL^2+L^4} \end{bmatrix}; \{w_e\}_4$$

$$= \left\{ \begin{array}{c} w_1 \\ \phi_1 \\ w_2 \\ \phi_2 \end{array} \right\} \tag{2.66}$$

In the light of Eq. (2.65), the assumed transverse displacement and rotation functions become:

$$w = \begin{bmatrix} 1 & x & x^2 & x^3 \end{bmatrix}\{a\} = \underline{\begin{bmatrix} 1 & x & x^2 & x^3 \end{bmatrix} [M_4]^{-1}}\{w_e\}_4 \tag{2.67}$$

$$\phi = \begin{bmatrix} 0 & 1 & 2x & 6g+3x^2 \end{bmatrix}\{a\}$$
$$= \underline{\begin{bmatrix} 0 & 1 & 2x & (6g+3x^2) \end{bmatrix} [M_4]^{-1}}\{w_e\}_4 \tag{2.68}$$

The necessary interpolation functions are obtained by evaluating the underlined matrix multiplication:

$$w = \underbrace{\left[\frac{(L-x)(Lx-2x^2+L^2(1+\varphi))}{L^3(1+\varphi)} \quad \frac{(L-x)x(-2x+L(2+\varphi))}{2L^2(1+\varphi)} \quad \frac{x(3Lx-2x^2+L^2\varphi)}{L^3(1+\varphi)} \quad -\frac{(L-x)x(2x+L\varphi)}{2L^2(1+\varphi)} \right]}_{[N_w]}\{w_e\}_4$$

$$\tag{2.69}$$

$$\phi = \underbrace{\left[\frac{6x(-L+x)}{L^3(1+\varphi)} \quad \frac{(L-x)(L-3x+L\varphi)}{L^2(1+\varphi)} \quad \frac{6(L-x)x}{L^3(1+\varphi)} \quad \frac{x(3x+L(-2+\varphi))}{L^2(1+\varphi)} \right]}_{[N_\phi]}\{w_e\}_4 \tag{2.70}$$

where $\psi = 12EI/kAGL^2$, while $[N_w]$ and $[N_\phi]$ represent the vector of interpolation functions for the transverse displacement and rotation of cross-section, respectively. Using $[N_w]$ and $[N_\phi]$, we can evaluate the non-zero components of the strain and rotation gradient tensors in terms of the following B matrices:

$$[B_\phi]_4 = \frac{d[N_\phi]}{dx}$$
$$= \left[-\frac{6(L-2x)}{L^3(1+\varphi)} \quad \frac{6x-L(4+\varphi)}{L^2(1+\varphi)} \quad \frac{6(L-2x)}{L^3(1+\varphi)} \quad \frac{6x+L(-2+\varphi)}{L^2(1+\varphi)} \right] \tag{2.71}$$

$$[B_\gamma]_4 = \left(-[N_\phi] + \frac{d[N_w]}{dx}\right)$$

$$= \left[-\frac{\varphi}{L+L\varphi} \quad -\frac{\varphi}{2+2\varphi} \quad \frac{\varphi}{L+L\varphi} \quad -\frac{\varphi}{2+2\varphi}\right] \tag{2.72}$$

$$[B_\chi]_4 = \frac{1}{2}\left(-\frac{d[N_\phi]}{dx} - \frac{d^2[N_w]}{dx^2}\right)$$

$$= \left[\frac{6(L-2x)}{L^3(1+\varphi)} \quad \frac{-6x+L(4+\varphi)}{L^2(1+\varphi)} \quad -\frac{6(L-2x)}{L^3(1+\varphi)} \quad \frac{6x+L(-2+\varphi)}{L^2(1+\varphi)}\right] \tag{2.73}$$

Therefore:

$$\{\varepsilon_{xx}\} = -z[B_\phi]_4\{w_e\}_4; \{\varepsilon_{xz}\}$$

$$= [B_\gamma]_4\{w_e\}_4; \{\chi_{xy}\}$$

$$= \frac{1}{2}[B_\chi]_4\{w_e\}_4 \tag{2.74}$$

$$\{\sigma_{xx}\} = -Ez[B_\phi]_4\{w_e\}_4; \{\sigma_{xz}\}$$

$$= Gk[B_\gamma]_4\{w_e\}_4; \{m_{xy}\}$$

$$= Gl^2[B_\chi]_4\{w_e\}_4 \tag{2.75}$$

Plugging Eqs. (2.74) and (2.75) into the total potential energy expression and then minimizing with respect to $\{w_e\}_4$ yields.

$$\frac{\partial \Pi}{\partial \{w_e\}_4^T} = 0$$

$$= \left(EI \int_0^{L_e} [B_\phi]_4^T [B_\phi]_4 \, dx + kGA \int_0^{L_e} [B_\gamma]_4^T [B_\gamma]_4 \, dx \right.$$

$$\left. + (Gl^2 A) \int_0^{L_e} [B_\chi]_4^T [B_\chi]_4 \, dx \right) \{w_e\}_4 - \{f_e\}_4 \tag{2.76}$$

Consequently, the elastostatic matrix equation of the 4 d.o.f microscale Timoshenko beam element is obtained as:

$$\{f_e\}_4 = [K_{4DOF}]\{w_e\}_4 \tag{2.77}$$

where

$$\{f_e\}_4 = \begin{Bmatrix} f_{z1} \\ m_1 \\ f_{z2} \\ m_2 \end{Bmatrix}; \; \{w_e\}_4 = \begin{Bmatrix} w_1 \\ \phi_1 \\ w_2 \\ \phi_2 \end{Bmatrix} \tag{2.78}$$

$$[K_{4DOF}] = [K_{bending}] + [K_{shear}] + [K_{mcst}] \tag{2.79}$$

$$[K_{bending}] = \begin{bmatrix} \frac{12EI}{L^2(1+\varphi)^2} & \frac{6EI}{L^2(1+\varphi)^2} & -\frac{12EI}{L^3(1+\varphi)^2} & \frac{6EI}{L^2(1+\varphi)^2} \\ \frac{6EI}{L^2(1+\varphi)^2} & \frac{EI\left(1+\frac{2}{(1+\varphi)^2}\right)}{L} & -\frac{6EI}{L^2(1+\varphi)^2} & -\frac{EI(-2+\varphi(2+\varphi))}{L(1+\varphi)^2} \\ -\frac{12EI}{L^3(1+\varphi)^2} & -\frac{6EI}{L^2(1+\varphi)^2} & \frac{12EI}{L^3(1+\varphi)^2} & -\frac{6EI}{L^2(1+\varphi)^2} \\ \frac{6EI}{L^2(1+\varphi)^2} & -\frac{EI(-2+\varphi(2+\varphi))}{L(1+\varphi)^2} & -\frac{6EI}{L^2(1+\varphi)^2} & \frac{EI\left(1+\frac{3}{(1+\varphi)^2}\right)}{L} \end{bmatrix} \tag{2.80}$$

$$[K_{shear}] = \begin{bmatrix} \frac{12EI\varphi}{L^3(1+\varphi)^2} & \frac{6EI\varphi}{L^2(1+\varphi)^2} & -\frac{12EI\varphi}{L^3(1+\varphi)^2} & \frac{6EI\varphi}{L^2(1+\varphi)^2} \\ \frac{6EI\varphi}{L^2(1+\varphi)^2} & \frac{6EI\varphi}{L(1+\varphi)^2} & -\frac{6EI\varphi}{L^2(1+\varphi)^2} & \frac{3EI\varphi}{L(1+\varphi)^2} \\ -\frac{12EI\varphi}{L^3(1+\varphi)^2} & -\frac{6EI\varphi}{L^2(1+\varphi)^2} & \frac{12EI\varphi}{L^3(1+\varphi)^2} & -\frac{6EI\varphi}{L^2(1+\varphi)^2} \\ \frac{6EI\varphi}{L^2(1+\varphi)^2} & \frac{3EI\varphi}{L(1+\varphi)^2} & -\frac{6EI\varphi}{L^2(1+\varphi)^2} & \frac{3EI\varphi}{L(1+\varphi)^2} \end{bmatrix} \tag{2.81}$$

$$[K_{mcst}] = \begin{bmatrix} \frac{12AG\ell^2}{L^3(1+\varphi)^2} & \frac{6AG\ell^2}{L^2(1+\varphi)^2} & -\frac{12AG\ell^2}{L^3(1+\varphi)^2} & \frac{6AG\ell^2}{L^2(1+\varphi)^2} \\ \frac{6AG\ell^2}{L^2(1+\varphi)^2} & \frac{AG\ell^2\left(1+\frac{3}{(1+\varphi)^2}\right)}{L} & -\frac{6AG\ell^2}{L^2(1+\varphi)^2} & -\frac{AG\ell^2(-2+2\varphi+\varphi^2)}{L^2(1+\varphi)^2} \\ -\frac{12AG\ell^2}{L^3(1+\varphi)^2} & -\frac{6AG\ell^2}{L^2(1+\varphi)^2} & \frac{12AG\ell^2}{L^3(1+\varphi)^2} & -\frac{6AG\ell^2}{L^2(1+\varphi)^2} \\ \frac{6AG\ell^2}{L^2(1+\varphi)^2} & -\frac{AG\ell^2(-2+2\varphi+\varphi^2)}{L^2(1+\varphi)^2} & -\frac{6AG\ell^2}{L^2(1+\varphi)^2} & \frac{AG\ell^2\left(1+\frac{3}{(1+\varphi)^2}\right)}{L} \end{bmatrix} \tag{2.82}$$

Section 2.4 details the implementation of the derived finite element model in the R programming language. For brevity sake, only the 4-d.o.f model is implemented.

2.4 R Functions for Static Analysis

A set of functions developed for this specific chapter are listed here. However, as mentioned in Chap. 1, a package named `microfiniteR` has been developed for this book. This package contains all the functions for this chapter and the subsequent ones. The package and the associated codes can be downloaded from the book's website or from https://github.com/khameelbm/microfiniteR. Its usage is demonstrated with examples in Sect. 2.5

Forming the stiffness matrix

```
FormStiffnessMTB<- function(youngmod, shearmod,
                            momentinertia, area,
                            shearfactor, poissonratio,
                            length, lengthscale){

  L <- length
  ele_DOF <- 4
  phi <- (12 * youngmod * momentinertia) / (shearfactor * area* shearmod * L^2)
  beta1 <- (area * shearmod * lengthscale^2) / ((1 + phi)^2 * L^3)
  beta2 <- (youngmod * momentinertia)/((1 + phi) * L^3)
  bendshear <- beta2 * matrix(c(12, 6*L, -12, 6*L,
                                6*L, (4+phi)*L^2, -6*L, (2-phi)*L^2,
                                -12, -6*L, 12, -6*L,
                                6*L, (2-phi)*L^2, -6*L,(4+phi)*L^2),
                                nrow <- ele_DOF,byrow=T)

  couplestress <- beta1 * matrix(c(12, 6*L, -12, 6*L,
                                   6*L, (4 + 2*phi + phi^2)*L^2,
                                   -6*L, -(-2 + 2*phi + phi^2)*L^2,
                                   -12, -6*L, 12, -6*L,
                                   6*L, -(-2 + 2*phi + phi^2)*L^2,
                                   -6*L, (4 + 2*phi + phi^2)*L^2),
                                   nrow=ele_DOF,byrow=T)
  totalstiffness <- bendshear + couplestress
  return(totalstiffness)
}
```

Forming the expanded matrix

```
ExpandStiffnessMTB <-  function(globa_dof, eMatrix, i, j) {

  r1 <- (1-1)+1
  r2 <- (i-1)+(i+1)
  r3 <- (j-2)+(j+1)
  r4 <- (j-2)+(j+2)

  mtbbigMatrix <- matrix(vector(l=globa_dof * globa_dof), nrow=globa_dof,
                         byrow=T)
  mtbbigMatrix[c(r1,r2,r3,r4), c(r1,r2,r3,r4)] <- eMatrix

  return (mtbbigMatrix)

}
```

Forming the reduced stiffness matrix and the reduced force vector

```
FormReducedStiffness <- function(globalK, knownforcenodes){

  reducedstiff <- globalK[c(knownforcenodes), (knownforcenodes)]
  return(reducedstiff)
}


FormReducedForce <- function(forcevector){

  reducedforcevector <- matrix(forcevector, ncol=1)
  return(reducedforcevector)
}
```

Functions for hinged and fixed boundary conditions

```
HingeNodes <- function(nodes){

  len=length(nodes)
  hinged_dof=vector(mode="numeric")
  for (k in 1:len){
    hinged_dof[k] <- (nodes[k]-1) + nodes[k]

  }

  return(hinged_dof)
}


FixNodes=function(nodes){
  len <- length(nodes)
  fixed_dof <- vector(mode="numeric")
  if (length(nodes)==1){
    fixed_dof[nodes] <- (nodes-1) + nodes
    fixed_dof[nodes+1] <- ((nodes-1) + (nodes+1))
    return(fixed_dof)
  } else {

    hinged_dof <- vector(mode="numeric")
    fixed_dof <- vector(mode="numeric")
    for (k in 1:len){
      hinged_dof[k] <- (nodes[k]-1) + nodes[k]
    }

    for (k in (1:len)){
      fixed_dof[k] <- (nodes[k]-1) + (nodes[k]+1)
      }
  }

  return(sort(c(hinged_dof, fixed_dof)))
}
```

Finding the unknown nodal degrees of freedom

```
FindNodalDOF <- function(reducedk, reducedforce){

  nodal_dof <- solve(reducedk, reducedforce)
  return(nodal_dof)
}
```

Finding the element forces

```
FindForcesMomentsMTB <- function(youngmod, shearmod,
                                 momentinertia, area,
                                 shearfactor, poissonratio,
                                 length,lengthscale, global_dof,
                                 i,j){

    ele_DOF <- 4
    r1  <-  (i - 1) + i;
    r2  <-  (i - 1) + (i + 1);
    r3  <-  (j - 2) + (j + 1);
    r4  <-  (j - 2) + (j + 2);
    nodaldisp <- global_dof[c(r1,r2,r3,r4)];
    L <- length
    phi <- (12 * youngmod * momentinertia) / (shearfactor * area* shearmod * L^2)
    beta1 <- (area * shearmod * lengthscale^2) / ((1 + phi)^2 * L^3)
    beta2 <- (youngmod * momentinertia)/((1 + phi) * L^3)

    bendshear <- beta2 * matrix(c(12, 6*L, -12, 6*L,
                                  6*L, (4+phi)*L^2, -6*L,
                                  (2-phi)*L^2,-12, -6*L,
                                  12, -6*L,6*L,
                                  (2-phi)*L^2, -6*L,(4+phi)*L^2),
                                  nrow <- ele_DOF,byrow=T)

    couplestress <- beta1 * matrix(c(12, 6*L, -12, 6*L,
                                     6*L, (4 + 2*phi + phi^2)*L^2,
                                     -6*L, -(-2 + 2*phi + phi^2)*L^2,
                                     -12, -6*L, 12, -6*L,
                                     6*L, -(-2 + 2*phi + phi^2)*L^2,
                                     -6*L, (4 + 2*phi + phi^2)*L^2),
                                         nrow=ele_DOF,byrow=T)
    totalstiffness <- bendshear + couplestress
    local_forcesmoments <- totalstiffness %*% matrix(nodaldisp,ncol=1)
    return(local_forcesmoments)
}
```

2.5 Using the Implemented Functions for Static Problems

Example 2.1 Cantilever microstructure-dependent beam

Problem Figure 2.4 shows a microscale cantilever beam with a tip load of $100\,\mu$N. Determine the deflection at the free end by using the following material and geometric properties [47]:
$$E = 1.44\,GPa;\ G = \frac{E}{2(1+v)};\ v = 0.38;\ k = \frac{5+5v}{6+5v};\ h = 17.6\,\mu\text{m};\ b = 20\,h;$$
$$I = \frac{bh^3}{12};\ A = bh;\ L = 12\,h;\ l = \left(\frac{1}{3}\right)h.$$

Solution

In this example, we will employ a one-element solution and then compare the computed results with the reported values in [47]. In practice, more than one element is required for most problems. That said, the choice of one element here hinges on the fact that for a cantilever beam, the only applied load (tip force at the right edge)

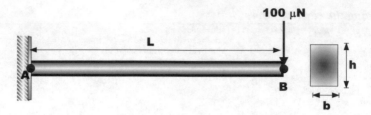

Fig. 2.4   A microscale cantilever beam with a tip load of 100 μN

and the constraint (at the left edge) coincide with the nodes of a two-node element.
As a result, a one-element solution provides an accurate agreement with the analytic
solution[10] [52]. Premised on this, points *A* and *B* become node 1 and 2, respectively.
The rest of the computations are now carried out using the developed R functions.

Step 1—Load the `microfiniteR` package, supply the material/geometric properties of the structure and declare the number of elements.

```
library(microfiniteR)
eE=1.44*(10^9)
pratio=0.38
eG=eE/(2*(1+pratio))
ek=(5+5*pratio)/(6+5*pratio)
h=17.6e-6
b=2*h
eI=(b*h^3)/12
eArea=b*h
eLen=(12*h)
eLenscale=(1/3)*h

numElement=1
```

Step 2—Form the element stiffness matrix using the function
`FormStiffnessMTB()`.

```
k1=FormStiffnessMTB(eE, eG, eI,eArea,ek,pratio,eLen,eLenscale);k1
```

The output of the above produces the element's stiffness matrix as:

```
> k1=FormStiffnessMTB(eE, eG, eI,eArea,ek,pratio,eLen,eLenscale);k1
              [,1]           [,2]           [,3]           [,4]
[1,]  42.272097581   4.463934e-03  -42.272097581   4.463934e-03
[2,]   0.004463934   6.331011e-07   -0.004463934   3.096817e-07
[3,] -42.272097581  -4.463934e-03   42.272097581  -4.463934e-03
[4,]   0.004463934   3.096817e-07   -0.004463934   6.331011e-07
```

Step 3—Find the total global degree of freedom and expand the stiffness matrix
accordingly using the function `ExpandStiffnessMTB()`.

---

[10]This is true for static problems. For dynamic problems, more elements are necessary to obtain
accurate results.

```
totalDOF=(numElement+1)*2
K1=ExpandStiffnessMTB(totalDOF,k1,1,2)
```

Step 4—Establish the global stiffness matrix.

```
globalK=K1;
globalK
```

Since we are using a single element, the global stiffness matrix will be the same size (4 × 4) as that obtained in step 2 as shown below:

```
> globalK=K1;globalK
              [,1]            [,2]            [,3]            [,4]
[1,]   42.272097581   4.463934e-03  -42.272097581   4.463934e-03
[2,]    0.004463934   6.331011e-07   -0.004463934   3.096817e-07
[3,]  -42.272097581  -4.463934e-03   42.272097581  -4.463934e-03
[4,]    0.004463934   3.096817e-07   -0.004463934   6.331011e-07
```

Step 5—Apply boundary condition(s) on the global stiffness matrix to obtain the reduced stiffness matrix and the reduced force vector.

For the boundary conditions, we note that the beam is fixed at node 1, which eliminates the vertical and rotational displacements at node 1. In what follows, we call three functions: FixNodes(), ExtractFreeRows(), and FormReducedStiffness().

- A call to FixNodes() eliminates the degrees of freedom at node 1.
- ExtractFreeRows() returns the free degrees of freedom.
- FormReducedStiffness() produces the reduced stiffness matrix (by eliminating the rows and columns, in the global stiffness matrix that are related to node 1).

To know more about any of these functions use the help function. For instance, to know more about the last function (type this in the RStudio console: ? FormReducedStiffness).

```
freeRows=ExtractFreeRows(totalDOF,c(FixNodes(1)));freeRows
reducedk=FormReducedStiffness(globalK,freeRows);
reducedk
```

Running the above produces:

```
> reducedk
              [,1]            [,2]
[1,]  42.272097581  -4.463934e-03
[2,]  -0.004463934   6.331011e-07
```

Our reduced stiffness matrix is of size 2 × 2, which indicates that there must be two known load, which are: $F_{2z} = 100\,\mu\mathrm{N}$, $M_2 = 0\,\mathrm{Nm}$. Hence:

```
reducedF=FormReducedForce(c(100e-6,0));reducedF
```

When executed, the above yields:

```
> reducedF=FormReducedForce(c(100e-6,0));reducedF
          [,1]
[1,] 1e-04
[2,] 0e+00
```

Step 6—Find the unknown nodal displacements.

```
unknown_dofs=FindNodalDOFs(reducedk,reducedF);
unknown_dofs
```

From the above, we determine the nodal displacements in *meter* and *radian* as:

```
> unknown_dofs
              [,1]
[1,] 9.261540e-06
[2,] 6.530221e-02
```

The computed result reveals that:

$$\left\{ \begin{array}{c} w_{2z} \\ \phi_2 \end{array} \right\} = \left\{ \begin{array}{c} 9.2615\,\mu\text{m} \\ 5.53022\,\text{rad} \end{array} \right\}$$

In order to allow for comparison with the non-dimensional result reported by Dehrouyeh-Semnani and Bahrami [47], the displacement value is further multiplied by $100/L$. This is done next:

```
nondimensionaldisp=(unknown_dofs[1,]*100)/(eLen);
nondimensionaldisp
```

where we have used the syntax `unknown_dofs[1,]` to extract the displacement value. The result is:

```
> nondimensionaldisp
[1] 4.385199
```

The computed non-dimensional displacement value matches that reported in Table 1[11] of the study by Dehrouyeh-Semnani and Bahrami [47].

As elaborated in Jones and Nenadic [53], although the cantilever configuration is very popular in MEMs deployment, other configurations are also typically used. In the next example we will consider the analysis of a simply-supported beam.

Example 2.2 Simply-supported microstructure-dependent beam

Problem Figure 2.5 shows a simply -supported microscale beam with a mid-span load of $100\,\mu\text{N}$. Determine the mid-point deflection of the beam. Use the following material and geometric properties [47]:

---

[11]Please note that page 140 of the reference contains other results, here we have only done the computation for the case of $L = 12\,h$, $l = 1/3h$.

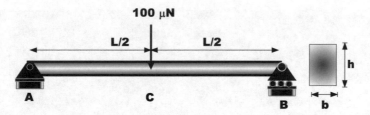

Fig. 2.5 A simply-supported microscale beam with a mid-point load of $100\,\mu N$

$$E = 1.44\,\text{GPa}; G = \frac{E}{2(1+v)}; v = 0.38; k = \frac{5+5v}{6+5v}; h = 17.6\,\mu\text{m}; b = 20\,h;$$
$$I = \frac{bh^3}{12}; A = bh; L = 12\,h; l = \left(\frac{1}{3}\right)h.$$

Solution

Here, a two-element solution is employed. As a result:

- points A, C and B become node 1, 2 and 3, respectively; and
- element 1 has nodes 1 & 2, while element 2 has nodes 2 & 3.

The rest of the computations for this problem are now carried out in a manner similar to Example 2.1.

Step 1—Load the microfiniteR package, supply the material/geometric properties of the structure and declare the number of elements.

```
library(microfiniteR)
eE=1.44*(10^9)
pratio=0.30
eG=eE/(2*(1+pratio))
ek=(5+5*pratio)/(6+5*pratio)
h=17.6e-6
b=2*h
eI=(b*h^3)/12
eArea=b*h
eLen=(12*h)/2
eLenscale=(1/3)*h


numElement=2
```

Step 2—Form the element matrix using the function FormStiffnessMTB().

```
k1=FormStiffnessMTB(eE, eG, eI, eArea, ek, pratio, eLen, eLenscale);k1
k2=FormStiffnessMTB(eE, eG, eI, eArea, ek, pratio, eLen, eLenscale);k2
```

This yields the two stiffness matrices:

```
> k1=FormStiffnessMTB(eE, eG, eI, eArea, ek, pratio, eLen, eLenscale);k1
              [,1]            [,2]           [,3]           [,4]
[1,]   311.53809880   1.644921e-02 -311.53809880   1.644921e-02
[2,]     0.01644921   1.191938e-06   -0.01644921   5.450990e-07
[3,] -311.53809880  -1.644921e-02  311.53809880  -1.644921e-02
[4,]     0.01644921   5.450990e-07   -0.01644921   1.191938e-06
> k2=FormStiffnessMTB(eE, eG, eI, eArea, ek, pratio, eLen, eLenscale);k2
              [,1]            [,2]           [,3]           [,4]
[1,]   311.53809880   1.644921e-02 -311.53809880   1.644921e-02
[2,]     0.01644921   1.191938e-06   -0.01644921   5.450990e-07
[3,] -311.53809880  -1.644921e-02  311.53809880  -1.644921e-02
[4,]     0.01644921   5.450990e-07   -0.01644921   1.191938e-06
```

Step 3—Find the total global degree of freedom and expand the stiffness matrices using the function `ExpandStiffnessMTB()`.

```
totalDOF=(numElement + 1)*2
K1=ExpandStiffnessMTB(totalDOF, k1, 1, 2)
K2=ExpandStiffnessMTB(totalDOF, k2, 2, 3)
```

Step 4—Sum up the expanded element matrices to establish the global stiffness matrix.

```
globalK=K1+K2
globalK
```

As seen in the next box, here the global stiffness matrix will be (6 × 6) compared to the 4 × 4 element stiffness matrix obtained in step 2.

```
> globalK
              [,1]            [,2]           [,3]           [,4]           [,5]           [,6]
[1,]   311.53809880   1.644921e-02 -311.53809880   1.644921e-02    0.00000000  0.000000e+00
[2,]     0.01644921   1.191938e-06   -0.01644921   5.450990e-07    0.00000000  0.000000e+00
[3,] -311.53809880  -1.644921e-02  623.07619761   0.000000e+00 -311.53809880  1.644921e-02
[4,]     0.01644921   5.450990e-07    0.00000000   2.383875e-06   -0.01644921  5.450990e-07
[5,]     0.00000000   0.000000e+00 -311.53809880  -1.644921e-02  311.53809880 -1.644921e-02
[6,]     0.00000000   0.000000e+00    0.01644921   5.450990e-07   -0.01644921  1.191938e-06
```

Step 5—Apply the boundary condition(s) on the global stiffness matrix to obtain the reduced stiffness matrix and supply information about the reduced force vector.

For this problem, nodes 1 and 3 are simply-supported, which means the vertical displacements at these two nodes are zero. Clearly, the slope at node 2 will also be zero and should be included in the set of applied boundary condition, but we will leave this out for the system to reveal. In the next line of code, we call three functions: `HingeNodes()`, `ExtractFreeRows()`, and `FormReducedStiffness()`.

Here `HingeNodes()` is called to eliminate the translational degree of freedom at nodes 1 and 3.

```
freeRows=ExtractFreeRows(totalDOF,c(HingeNodes(1),HingeNodes(3)))
reducedk=FormReducedStiffness(globalK,freeRows);reducedk
```

Running the above produces:

```
> reducedk=FormReducedStiffness(globalK, freeRows); reducedk
          [,1]          [,2]          [,3]          [,4]
[1,]  1.191938e-06  -0.01644921  5.450990e-07  0.000000e+00
[2,] -1.644921e-02 623.07619761  0.000000e+00  1.644921e-02
[3,]  5.450990e-07   0.00000000  2.383875e-06  5.450990e-07
[4,]  0.000000e+00   0.01644921  5.450990e-07  1.191938e-06
```

Our reduced stiffness matrix is of size 4×4, which indicates that there must be four known loads, which are: $M_1 = 0$, $F_{2z} = 100\,\mu\text{N}$, $M_2 = 0\,\text{Nm}$, and $M_3 = 0\,\text{Nm}$. Hence:

```
reducedF=FormReducedForce(c(0,100e-6,0,0))
```

Step 6—Find the unknown nodal displacements.

```
unknown_dofs=FindNodalDOFs(reducedk, reducedF);
unknown_dofs
```

From the above, we determine the nodal displacements in *meter* and *radian* as:

```
> unknown_dofs
            [,1]
[1,]  8.162776e-03
[2,]  5.914886e-07
[3,] -1.038278e-18
[4,] -8.162776e-03
```

From the output result, the computed values of the non-trivial displacements at nodes 1, 2 and 3 are[12]:

$$\left\{ \begin{array}{c} \phi_1 \\ w_{2z} \\ \phi_2 \\ \phi_3 \end{array} \right\} = \left\{ \begin{array}{c} 8.162776\,\text{rad} \\ 0.5914886\,\mu\text{m} \\ -1.03 \times 10^{-18} \approx 0 \\ -8.162776\,\text{rad} \end{array} \right\}$$

Again, to allow for comparison with the non-dimensional reference results, the displacement value is further multiplied by $100/L$. This is done next:

```
nondimensionaldisp=(unknown_dofs[2,]*100)/(2*eLen);
nondimensionaldisp
```

where we have used the syntax `unknown_dofs[2,]` to extract the displacement value at node 2. The result is:

```
> nondimensionaldisp
[1] 0.2800609
```

The output value matches that reported in Table 4[13] of Ref. [47].

---

[12] As expected the value of the slope at node 2 is approximately zero.

[13] Please note that page 140 of the reference contains other results, here we have only done the computation for the case of $L = 12\,h$, $l = 1/3h$.

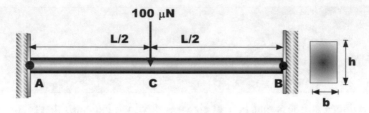

Fig. 2.6 A double clamped microscale beam with a mid-point load of 100 μN

Example 2.3 Double clamped microstructure-dependent beam

Problem Figure 2.6 shows a microscale clamped-clamped beam with a mid-span load of 100 μN. Determine the deflection at the mid-point of the beam. Use the following material and geometric properties [47]:

$E = 1.44\,\text{GPa}; G = \frac{E}{2(1+v)}; v = 0.38; k = \frac{5+5v}{6+5v}; h = 17.6\,\mu\text{m}; b = 20\,h;$

$I = \frac{bh^3}{12}; A = bh; L = 12\,h; l = \left(\frac{1}{3}\right)h.$

Solution

This problem also requires a two-element solution. Thus, the configuration and connectivity of the elements and the nodes is similar to that of Example 2.2. Based on this, the step by step computations for this problem have been compiled into a single snippet of code shown below. As mentioned in Chap. 1, any line that is preceded by '#' is a comment and will be ignored by the R compiler.

```r
# Load library and supply material/geometric properties
library(microfiniteR)
eE=1.44*(10^9)
pratio=0.38
eG=eE/(2*(1+pratio))
ek=(5+5*pratio)/(6+5*pratio)
h=17.6e-6
b=2*h
eI=(b*h^3)/12
eArea=b*h
eLen=(12*h)/2
eLenscale=(1/3)*h

numElement=2

# Form the element matrix using the function FormStiffnessMTB().
k1=FormStiffnessMTB(eE, eG, eI, eArea, ek, pratio, eLen, eLenscale);k1
k2=FormStiffnessMTB(eE, eG, eI, eArea, ek, pratio, eLen, eLenscale);k2

# Find the total global degree of freedom and expand the stiffness
# matrix accordingly using the function ExpandStiffnessMTB().
totalDOF=(numElement + 1)*2
K1=ExpandStiffnessMTB(totalDOF, k1, 1, 2)
K2=ExpandStiffnessMTB(totalDOF, k2, 2, 3)

# Establish the global stiffness matrix
globalK=K1+K2
globalK

# Apply the boundary condition(s) on the global stiffness
# matrix to obtain the reduced stiffness matrix and
# the reduced force vector.
freeRows=ExtractFreeRows(totalDOF,c(FixNodes(c(1,3))))
reducedk=FormReducedStiffness(globalK,freeRows);reducedk

reducedF=FormReducedForce(c(100e-6,0))

# Find the unknown nodal displacements.
unknown_dofs=FindNodalDOFs(reducedk,reducedF);
unknown_dofs

# Analytical validation for the mid-point deflection
# Note that eLen is multiplied by 2 here because we used 2 elements.
nondimensionaldisp=(unknown_dofs[1,]*100)/(2*eLen);
nondimensionaldisp
```

The combined outputs of the preceding code snippet is given next:

```
> k1=FormStiffnessMTB(eE, eG, eI, eArea, ek, pratio, eLen, eLenscale);k1
              [,1]          [,2]          [,3]          [,4]
[1,]   311.53809880   1.644921e-02  -311.53809880   1.644921e-02
[2,]     0.01644921   1.191938e-06    -0.01644921   5.450990e-07
[3,]  -311.53809880  -1.644921e-02   311.53809880  -1.644921e-02
[4,]     0.01644921   5.450990e-07    -0.01644921   1.191938e-06
> k2=FormStiffnessMTB(eE, eG, eI, eArea, ek, pratio, eLen, eLenscale);k2
              [,1]          [,2]          [,3]          [,4]
[1,]   311.53809880   1.644921e-02  -311.53809880   1.644921e-02
[2,]     0.01644921   1.191938e-06    -0.01644921   5.450990e-07
[3,]  -311.53809880  -1.644921e-02   311.53809880  -1.644921e-02
[4,]     0.01644921   5.450990e-07    -0.01644921   1.191938e-06
>


> globalK
              [,1]          [,2]          [,3]          [,4]          [,5]          [,6]
[1,]   311.53809880   1.644921e-02  -311.53809880   1.644921e-02   0.00000000   0.000000e+00
[2,]     0.01644921   1.191938e-06    -0.01644921   5.450990e-07   0.00000000   0.000000e+00
[3,]  -311.53809880  -1.644921e-02   623.07619761   0.000000e+00  -311.53809880   1.644921e-02
[4,]     0.01644921   5.450990e-07    0.00000000   2.383875e-06    -0.01644921   5.450990e-07
[5,]     0.00000000   0.000000e+00  -311.53809880  -1.644921e-02   311.53809880  -1.644921e-02
[6,]     0.00000000   0.000000e+00    0.01644921   5.450990e-07    -0.01644921   1.191938e-06
>


> reducedk=FormReducedStiffness(globalK,freeRows);reducedk
        [,1]           [,2]
[1,] 623.0762 0.000000e+00
[2,]   0.0000 2.383875e-06
>


> unknown_dofs=FindNodalDOFs(reducedk,reducedF);
> unknown_dofs
            [,1]
[1,] 1.60494e-07
[2,] 0.00000e+00
>


> nondimensionaldisp
[1] 0.07599148
```

The final output represents the non-dimensional mid-point displacement value (0.07599), which is close to that reported in Table 7 of [47].

2.6 Summary

In this chapter, the governing equations of a microstructure-dependent microscale beam, based on the assumptions of the Timoshenko beam theory and the framework of the modified couple stress theory, are derived. The derived governing equations are employed to develop a locking-free microscale Timoshenko beam (MTB) element, using a reduced-order model with 4 degrees of freedom. A set of functions, based on the R programming language, are provided to facilitate computations with the developed finite element model. Using three examples, we introduced the usage of microfiniteR (an R package), which houses the functions (and their documentations) developed in this book.

References

1. J.F. Doyle, *Static and Dynamic Analysis of Structures: With an Emphasis on Mechanics and Computer Matrix Methods* (Springer, Netherlands, 1991)
2. M.H. Kahrobaiyan, M. Asghari, M.T. Ahmadian, A strain gradient Timoshenko beam element: Application to MEMS. Acta Mech. **226**, 505–525 (2015)
3. R.A. Coutu Jr., P.E. Kladitis, L. Starman, J.R. Reid, A comparison of micro-switch analytic, finite element, and experimental results. Sens. Actuators, A **115**, 252–258 (2004)
4. V. Kaajakari, *Practical MEMS: Small Gear Pub.* (2009)
5. S. Khakalo, V. Balobanov, J. Niiranen. Modelling size-dependent bending, buckling and vibrations of 2D triangular lattices by strain gradient elasticity models: Applications to sandwich beams and auxetics. Int. J. Eng. Sci. **127**, 33–52 (2018)
6. B.R. Goncalves, A. Karttunen, J. Romanoff, J. Reddy, Buckling and free vibration of shear-flexible sandwich beams using a couple-stress-based finite element. Compos. Struct. **165**, 233–241 (2017)
7. A.T. Karttunen, J. Romanoff, J. Reddy, Exact microstructure-dependent Timoshenko beam element. Int. J. Mech. Sci. **111**, 35–42 (2016)
8. C.L. Dym, I.H. Shames, *Solid mechanics: A variational approach*, Augmented edn. (Springer, New York, 2013)
9. R. Mindlin, H. Tiersten, Effects of couple-stresses in linear elasticity. Arch. Ration. Mech. Anal. **11**, 415–448 (1962)
10. S. Wulfinghoff, E. Bayerschen, T. Böhlke, Micromechanical simulation of the Hall-Petch effect with a crystal gradient theory including a grain boundary yield criterion. PAMM **13**, 15–18 (2013)
11. W.B. Anderson, R.S. Lakes, Size effects due to Cosserat elasticity and surface damage in closed-cell polymethacrylimide foam. J. Mater. Sci. **29**, 6413–6419 (1994)
12. R. Lakes, Size effects and micromechanics of a porous solid. J. Mater. Sci. **18**, 2572–2580 (1983)
13. P.R. Onck, E.W. Andrews, L.J. Gibson, Size effects in ductile cellular solids. Part I: modeling. Int. J. Mech. Sci. **43**, 681–699 (2001)
14. P. Giovine, L. Margheriti, M.P. Speciale, On wave propagation in porous media with strain gradient effects. Comput. Math. Appl. **55**, 307–318 (2008)
15. A. Kelly, *Precipitation hardening* (Pergamon Press, 1963)
16. R. Ebeling, M.F. Ashby, Dispersion hardening of copper single crystals. Philos. Mag. **13**, 805–834 (1966)
17. M.F. Ashby, The deformation of plastically non-homogeneous materials. Philos. Mag. **21**, 399–424 (1970)

18. J.W. Hutchinson, Plasticity at the micron scale. Int. J. Solids Struct. **37**, 225–238 (2000)
19. D.C.C. Lam, A.C.M. Chong, Indentation model and strain gradient plasticity law for glassy polymers. J. Mater. Res. **14**, 3784–3788 (1999)
20. H. Askes, E.C. Aifantis, Gradient elasticity in statics and dynamics: An overview of formulations, length scale identification procedures, finite element implementations and new results. Int. J. Solids Struct. **48**, 1962–1990 (2011)
21. J. Dyszlewicz, *Micropolar Theory of Elasticity* (Springer, Heidelberg, 2004)
22. H.-T. Thai, T.P. Vo, T.-K. Nguyen, S.-E. Kim, A review of continuum mechanics models for size-dependent analysis of beams and plates. Compos. Struct. **177**, 196–219 (2017)
23. S. Khakalo, J. Niiranen, Form II of Mindlin's second strain gradient theory of elasticity with a simplification: For materials and structures from nano- to macro-scales. Eur. J. Mech. A/Solids **71**, 292–319 (2018)
24. W. Voigt, *Theoretische studien über die elasticitätsverhältnisse der krystalle*: Königliche Gesellschaft der Wissenschaften zu Göttingen (1887)
25. E. Cosserat, F. Cosserat, Théorie des corps déformables (1909)
26. R.D. Mindlin, H.F. Tiersten, Effects of couple-stresses in linear elasticity. Arch. Ration. Mech. Anal. **11**, 415–448 (1962)
27. R.A. Toupin, Elastic materials with couple-stresses. Arch. Ration. Mech. Anal. **11**, 385–414 (1962)
28. W. Koiter, Couple-stress in the theory of elasticity, in *Proceedings of the K. Ned. Akad. Wet* (1964), pp. 17–44
29. A.C. Eringen, A unified theory of thermomechanical materials. Int. J. Eng. Sci. **4**, 179–202 (1966)
30. A.C. Eringen, E.S. Suhubi, Nonlinear theory of simple micro-elastic solids—I. Int. J. Eng. Sci. **2**, 189–203 (1964)
31. R.D. Mindlin, Micro-structure in linear elasticity. Arch. Ration. Mech. Anal. **16**, 51–78 (1964)
32. R.D. Mindlin, N.N. Eshel, On first strain-gradient theories in linear elasticity. Int. J. Solids Struct. **4**, 109–124 (1968)
33. R.D. Mindlin, Second gradient of strain and surface-tension in linear elasticity. Int. J. Solids Struct. **1**, 417–438 (1965)
34. K.B. Mustapha, D. Ruan, Size-dependent axial dynamics of magnetically-sensitive strain gradient microbars with end attachments. Int. J. Mech. Sci. **94–95**, 96–110 (2015)
35. K.B. Mustapha, B.T. Wong, Torsional frequency analyses of microtubules with end attachments. ZAMM-J. Appl. Math. Mech./Zeitschrift für Angewandte Mathematik und Mechanik **96**, 824–842 (2016)
36. D.C.C. Lam, F. Yang, A.C.M. Chong, J. Wang, P. Tong, Experiments and theory in strain gradient elasticity. J. Mech. Phys. Solids **51**, 1477–1508 (2003)
37. F. Yang, A.C.M. Chong, D.C.C. Lam, P. Tong, Couple stress based strain gradient theory for elasticity. Int. J. Solids Struct. **39**, 2731–2743 (2002)
38. J.N. Reddy, Microstructure-dependent couple stress theories of functionally graded beams. J. Mech. Phys. Solids **59**, 2382–2399 (2011)
39. H.M. Ma, X.L. Gao, J.N. Reddy, A microstructure-dependent Timoshenko beam model based on a modified couple stress theory. J. Mech. Phys. Solids **56**, 3379–3391 (2008)
40. A.C.M. Chong, F. Yang, D.C.C. Lam, P. Tong, Torsion and bending of micron-scaled structures. J. Mater. Res. **16**, 1052–1058 (2001)
41. A.H. Nayfeh, P.F. Pai, *Linear and Nonlinear Structural Mechanics* (Wiley, 2008)
42. J.R. Barber, *Elasticity* (Springer, Heidelberg, 2002)
43. E.B. Magrab, *Vibrations of Elastic Systems: With Applications to MEMS and NEMS*, vol 184 (Springer Science & Business Media, 2012)
44. M.H. Kahrobaiyan, M. Asghari, M.T. Ahmadian, A Timoshenko beam element based on the modified couple stress theory. Int. J. Mech. Sci. **79**, 75–83 (2014)
45. I. Babuska, B.A. Szabo, I.N. Katz, The p-version of the finite element method. SIAM J. Numer. Anal. **18**, 515–545 (1981)

46. T. Apel, J.M. Melenk, Interpolation and quasi-Interpolation in h- and hp-version finite element spaces. *Encyclopedia of Computational Mechanics*, 2nd edn. (2017), pp. 1–33
47. A.M. Dehrouyeh-Semnani, A. Bahrami, On size-dependent Timoshenko beam element based on modified couple stress theory. Int. J. Eng. Sci. **107**, 134–148 (2016)
48. A. Arbind, J. Reddy, Nonlinear analysis of functionally graded microstructure-dependent beams. Compos. Struct. **98**, 272–281 (2013)
49. C. Liebold, W.H. Müller, Comparison of gradient elasticity models for the bending of micro-materials. Comput. Mater. Sci. **116**, 52–61 (2016)
50. Z. Li, Y. He, J. Lei, S. Guo, D. Liu, L. Wang, A standard experimental method for determining the material length scale based on modified couple stress theory. Int. J. Mech. Sci. **141**, 198–205 (2018)
51. R. Narayanaswami, H. Adelman, Inclusion of transverse shear deformation in finite element displacement formulations. AIAA J. **12**, 1613–1614 (1974)
52. A. Öchsner, *Computational Statics and Dynamics: An Introduction Based on the Finite Element Method* (Springer, Singapore, 2016)
53. T.B. Jones, N.G. Nenadic, *Electromechanics and MEMS* (Cambridge University Press, 2013)

Chapter 3
Vibration and Buckling of Microstructure-Dependent Timoshenko MicroBeams and Finite Element Implementations with R

Abstract The free oscillatory and elastic instability behaviours of microstructure-dependent beams are considered in this chapter. The equations governing these behaviours are established in accordance with the Timoshenko beam theory within the framework of the modified couple stress theory. For the free vibration analysis, the coupled partial differential equations exhibit space-time dependency, while only space dependency exists for buckling analyses. In each case, the effect of couple stress is considered under different boundary conditions using the finite element method. The variations of the natural frequencies (in the case of a free vibration analysis) and the critical buckling loads (for the buckling analysis) with changes in the small-scale parameter are evaluated and presented.

3.1 Introduction

In the study of microstructure-dependent beams in Chap. 2, we assumed the structures to be under the influence of time-independent transverse loads. Here, we examine the case when microscale beams are subjected to: (a) time-dependent loads—leading to dynamics effect and; (b) axial compressive loads, capable of leading to the loss of the structure's bending stiffness (buckling effect). Indisputably, a considerable body of work exists on the vibration and buckling of skeletal structural members (beams, plates, shells etc.). A compendium of analytical approaches to vibration analysis of structures under different types of loading and boundary conditions are detailed in many excellent references (e.g. [1–3]). On the other hand, the finite element treatment of vibration problems of classical beams and plates can be found in [4–6]. An excellent treatment of the various forms of classical Timoshenko beams can also be found in Reddy [7].

The practical requirement to determine the vibration and buckling behaviours of microscale structures exists in many applications. For instance, in microscale power generators as well as vibration-based MEMs applications, the determination of the resonant frequencies constitutes an important phase of the design flow [8, 9]. Indeed, the operating frequency range of a number of sensors (e.g. accelerometers, pressure

K. B. Mustapha, *R for Finite Element Analyses of Size-dependent Microscale Structures*, SpringerBriefs in Computational Mechanics, https://doi.org/10.1007/978-981-13-7014-4_3

sensors, microphones etc.) is influenced by the resonant frequencies of their internal vibrating elements [10], and certain actuators rely on judicious use of the resonant frequencies of the internal vibrating elements to achieve larger deflections. Further, the characterization of the buckling behaviour is of significant interest in bi-stable actuators [11].

The reality that the dimensions of the sensing and actuating elements in the afore-mentioned devices and many others are in the microscale domain enforces the need to account for the size-dependency that follows. One of the earliest simplified the-oretical approach to account for the size-effect in the analyses of microscale beams was given by Kong et al. [12]. In the study, an analytical solution to a model based on microstructure-dependent Euler-Bernoulli beams was reported. In this study, the size-effect was accounted for in the form of a single material length scale. Further, the study indicates that when the size-effect is accounted for, the natural frequencies of the beam show a large deviation from that obtained from the classical continuum theory. Shortly after the work by Kong et al. [12], the model and analytical solutions to the vibration of microstructure-dependent Timoshenko beam was presented by Ma et al. [13]. Hundreds of other scholarly works have since appeared offering different levels of insight into the consideration of size-effect in the vibration and buckling of microscale structures under various types of mechanical and electromechanical loading conditions [14–17].

Yet, many of the existing studies focus on analytical solutions that are limited to certain boundary conditions. The aim of this chapter is to present the finite element models for the free vibration and buckling analyses of microstructure-dependent Timoshenko beams. In Sect. 3.2, the distributed mathematical model of microscale beams with size-dependent properties will be derived based on the modified couple stress theory. Next, the finite element formulation of the presented model follows in Sect. 3.3. The collection of **R** functions implemented to support the numerical evaluation of the model will be in Sect. 3.4, and this is followed by a summary of the chapter.

3.2 Governing Differential Equations

As already indicated in the preceding chapter, the governing equations to be estab-lished for both free vibration and buckling analyses are based on the framework of the modified couple stress theory (MCST). However, for the purpose of dealing with dynamic effects, the extended Hamilton's principle (another variational technique) will be used to derive the requisite equations. The principle requires a knowledge of the kinetic and potential energy functions, and it demands that the true dynamic tra-jectory of the structure's motion in the configuration space between the time interval t_1 and t_2 imposes a stationary value on the integral of the energy functional I in the manner shown in Eq. (3.1) [18]:

$$I = \int_{t_1}^{t_2} \mathcal{L} dt; \quad \mathcal{L} = \Pi_K - \Pi_P \tag{3.1}$$

where Π_K is the kinetic energy (which helps to account for the structure's inertia), while Π_P denotes the total potential energy. In turn, the total potential energy consists of the strain energy Π_s and the effect of the external work done Π_W.

Given a mass density ρ and a velocity field V arising from an admissible displacement field u following an oscillatory motion induced by a transverse load, the kinetic energy can be obtained from:

$$\Pi_K = \frac{1}{2} \int_\Omega \rho V^2 d\Omega = \frac{1}{2} \int_0^L \int_A \rho V^2 dA dx \tag{3.2}$$

In Chap. 2, the size-dependent strain energy (Π_s) stored in a deformed isotropic linear elastic body occupying a volume Ω was given as [19, 20]:

$$\Pi_s = \frac{1}{2} \int_\Omega \left(\sigma_{ij} \varepsilon_{ij} + m_{ij} \chi_{ij} \right) d\Omega \tag{3.3}$$

3.2.1 Displacement Field

The reduced displacement field for the Timoshenko beam is re-written to account for the time dependency of the motion [21]:

$$u_1(x, y, z, t) = -z\phi(x, t); u_2(x, y, z, t) = 0; u_3(x, y, z, t) = w(x, t) \tag{3.4}$$

where u_1 is the axial displacement induced by bending, while the lateral displacements along the y-axis and z-axis are denoted by u_2 and u_3, respectively. The cross-sectional rotation (with respect to a transverse normal line) is denoted by ϕ. Further, w is the total transverse displacement due to shear and bending deformations. To obtain the full expression for the strain energy, we will employ the following earlier obtained non-zero components of the strain, the rotation gradient, the stress and the couple stress tensors:

$$\varepsilon_{xx} = \frac{\partial u_1}{\partial x} = -z\frac{\partial \phi}{\partial x};$$
$$\varepsilon_{xz} = \varepsilon_{zx} = \frac{\gamma_{xz}}{2} = \frac{1}{2}\left(\frac{\partial u_1}{\partial z} + \frac{\partial u_3}{\partial x} \right) = \frac{1}{2}\left(-\phi + \frac{\partial w}{\partial x} \right) \tag{3.5}$$

$$\chi_{xy} = \chi_{yx} = \frac{1}{4}\left(-\frac{\partial\phi}{\partial x} - \frac{\partial^2 w}{\partial x^2}\right) \tag{3.6}$$

$$\sigma_{xx} = E\varepsilon_{xx} = -Ez\frac{\partial\phi}{\partial x}; \sigma_{xz} = 2G\varepsilon_{xz} = \mu\left(-\phi + \frac{\partial w}{\partial x}\right) \tag{3.7}$$

$$m_{xy} = m_{yx} = 2Gl^2\chi_{xy} = \frac{Gl^2}{2}\left(-\frac{\partial\phi}{\partial x} - \frac{\partial^2 w}{\partial x^2}\right) \tag{3.8}$$

where l, E and G represent the material length scale, the Young's modulus and the shear modulus, respectively. Substituting the above components of the tensors in the strain energy density function of the microscale Timoshenko beam, we have:

$$\Pi_s = \frac{1}{2}\int_0^L EI\left(\frac{\partial\phi}{\partial x}\right)^2 dx + \frac{1}{2}\int_0^L GAk\left(-\phi + \frac{\partial w}{\partial x}\right)^2 dx$$

$$+ \frac{1}{2}\int_0^L \frac{Gl^2A}{4}\left(-\frac{\partial\phi}{\partial x} - \frac{\partial^2 w}{\partial x^2}\right)^2 dx \tag{3.9}$$

The parameters in Eq. (3.9) have the same meaning as those in Chap. 2. The total work done by the distributed load $q(x)$ acting at the centroid of the beam between the end points $[0, L]$ and an axial compressive force $p(x, t)$ can be written as [22]:

$$\Pi_W = -\int_0^L q(x, t)wdx + \frac{1}{2}\int_0^L p(x, t)\left(\frac{\partial w}{\partial x}\right)^2 dx \tag{3.10}$$

It then follows from Eqs. (3.9) and (3.10) that the total potential energy takes the form (isotropic material properties):

$$\Pi_P = \Pi_s + \Pi_W$$

$$= \frac{1}{2}\int_0^L \left[EI\left(\frac{\partial\phi}{\partial x}\right)^2 + GAk\left(-\phi + \frac{\partial w}{\partial x}\right)^2 + \frac{Gl^2A}{4}\left(-\frac{\partial\phi}{\partial x} - \frac{\partial^2 w}{\partial x^2}\right)^2\right.$$

$$\left. - q(x, t)w(x, t) + p(x, t)\left(\frac{\partial w}{\partial x}\right)^2\right]dx \tag{3.11}$$

The kinetic energy is obtained as follows:

$$\Pi_K = \frac{1}{2} \int_0^L \int_A \rho \left[\left(\frac{\partial u_1}{\partial t} \right)^2 + \left(\frac{\partial u_2}{\partial t} \right)^2 + \left(\frac{\partial u_3}{\partial t} \right)^2 \right] dA dx$$

$$= \frac{1}{2} \int_0^L \int_A \rho \left[\left(-z \frac{\partial \phi}{\partial t} \right)^2 + \left(\frac{\partial w}{\partial t} \right)^2 \right] dA dx$$

$$= \frac{1}{2} \int_0^L \rho A \left(\frac{\partial w}{\partial t} \right)^2 + \rho I \left(\frac{\partial \phi}{\partial t} \right)^2 dx \tag{3.12}$$

By combining Eqs. (3.11) and (3.12), we can re-write (3.1) as:

$$I = \int_{t_1}^{t_2} \left\{ \frac{1}{2} \int_0^L \left[\rho A \left(\frac{\partial w}{\partial t} \right)^2 + \rho I \left(\frac{\partial \phi}{\partial t} \right)^2 \right] dx \right.$$

$$- \left[EI \left(\frac{\partial \phi}{\partial x} \right)^2 + G A k \left(-\phi + \frac{\partial w}{\partial x} \right)^2 + \frac{G I^2 A}{4} \left(-\frac{\partial \phi}{\partial x} - \frac{\partial^2 w}{\partial x^2} \right)^2 \right]$$

$$\left. - q(x,t)w(x,t) + p(x,t) \left(\frac{\partial w}{\partial x} \right)^2 \right] dx \right\} dt \tag{3.13}$$

which, in terms of a functional F, can be re-cast as[1]:

$$I = \int_{t_2}^{t_2} \left\{ \int_0^L F(x, t, \phi, w, w_t, \phi_t, \phi_x, w_x, w_{xx}) dx \right\} dt \tag{3.14}$$

where x and t are the independent variables.

3.2.2 The Extended Hamilton's Principle

We now need to obtain the extremum of $I(w, \phi)$, within the time interval t_2 and t_1 as stipulated by the extended Hamilton's principle. A way around this requirement is to obtain the variational derivative of I (i.e. δI). In consonance with [1, 21], obtaining the variational derivative of $I(w, \phi)$ can be achieved by defining two families of trial functions:

---

[1]The subscript notations have been used to denote differential operators such that: $\phi_t = \frac{\partial \phi}{\partial t}$; $\phi_x = \frac{\partial \phi}{\partial x}$; $w_t = \frac{\partial w}{\partial t}$; $w_x = \frac{\partial w}{\partial x}$; $w_{xx} = \frac{\partial^2 w}{\partial x^2}$.

$$\bar{\phi}(x, t) = \phi(x, t) + \in \alpha_1(x, t) \tag{3.15}$$

$$\bar{w}(x, t) = w(x, t) + \in \alpha_2(x, t) \tag{3.16}$$

where $\bar{\phi}(x, t)$ and $\bar{w}(x, t)$ represent the variations in the original field variables (ϕ and w), while $\in$ is a small parameter. Besides, α_1 and α_2 vanish at the boundaries, and they belong to members of differentiable continuous functions. Connecting Eqs. (3.15) and (3.16) with (3.13), leads to:

$$\bar{I}(\in) = \int_{t_1}^{t_2} \left\{ \int_0^L F\left(x, t, \bar{\phi}, \bar{w}, \bar{w}_t, \bar{\phi}_t, \bar{\phi}_x, \bar{w}_x, \bar{w}_{xx}\right) dx \right\} dt \tag{3.17}$$

A necessary condition to obtain the extremum of $\bar{I}(\in)$ then turns out to require:

$$\left.\frac{d\bar{I}}{d\in}\right|_{\in=0} = \frac{d}{d\in}\bar{I}(\phi + \in \alpha_1, w + \in \alpha_2)\bigg|_{\in=0} = 0 \tag{3.18}$$

Mapping a differential operator to $\bar{I}$ results in:

$$\frac{d\bar{I}}{d\in} = \int_{t_1}^{t_2} \left\{ \int_0^L \left(\begin{array}{l} \dfrac{\partial F}{\partial \bar{\phi}}\dfrac{\partial \bar{\phi}}{\partial \in} + \dfrac{\partial F}{\partial \bar{w}}\dfrac{\partial \bar{w}}{\partial \in} + \dfrac{\partial F}{\partial \bar{\phi}_t}\dfrac{\partial \bar{\phi}_t}{\partial \in} + \dfrac{\partial F}{\partial \bar{w}_t}\dfrac{\partial \bar{w}_t}{\partial \in} \\[3mm] + \dfrac{\partial F}{\partial \bar{\phi}_x}\dfrac{\partial \bar{\phi}_x}{\partial \in} + \dfrac{\partial F}{\partial \bar{w}_x}\dfrac{\partial \bar{w}_x}{\partial \in} + \dfrac{\partial F}{\partial \bar{w}_{xx}}\dfrac{\partial \bar{w}_{xx}}{\partial \in} \end{array} \right) dx \right\} dt \tag{3.19}$$

The trial functions ($\bar{\phi}$ and $\bar{w}$) will qualify as minimizing functions if $\in = 0$, therefore:

$$\left.\frac{d\bar{I}}{d\in}\right|_{\in=0} = \int_{t_1}^{t_2} \left\{ \int_0^L \left(\begin{array}{l} \dfrac{\partial F}{\partial \phi}\alpha_1 + \dfrac{\partial F}{\partial w}\alpha_2 + \underline{\dfrac{\partial F}{\partial \phi_t}\alpha_{1t}} + \underline{\dfrac{\partial F}{\partial w_t}\alpha_{2t}} \\[3mm] + \underline{\dfrac{\partial F}{\partial \phi_x}\alpha_{1x}} + \underline{\dfrac{\partial F}{\partial w_x}\alpha_{2x}} + \underline{\dfrac{\partial F}{\partial w_{xx}}\alpha_{2xx}} \end{array} \right) dx \right\} dt \tag{3.20}$$

A series of integration by part is carried out on the underlined terms in Eq. (3.20) to retrieve the Euler-Lagrange equations of the system as:

$$\frac{\partial F}{\partial \phi} - \frac{d}{dt}\frac{\partial F}{\partial \phi_t} - \frac{d}{dx}\frac{\partial F}{\partial \phi_x} = 0 \tag{3.21}$$

$$\frac{\partial F}{\partial w} - \frac{d}{dt}\frac{\partial F}{\partial w_t} - \frac{d}{dx}\frac{\partial F}{\partial w_x} + \frac{d^2}{dx^2}\frac{\partial F}{\partial w_{xx}} = 0 \tag{3.22}$$

where

$$\frac{\partial F}{\partial \phi} = AGk\left(-\phi + \frac{\partial w}{\partial x}\right); \quad \frac{\partial F}{\partial \phi_t} = \rho I \frac{\partial \phi}{\partial t};$$

$$\frac{\partial F}{\partial \phi_x} = -EI\frac{\partial \phi}{\partial x} - \frac{1}{4}GAl^2\left(\frac{\partial \phi}{\partial x} + \frac{\partial^2 w}{\partial x^2}\right)$$

$$\frac{\partial F}{\partial w} = -q; \quad \frac{\partial F}{\partial w_t} = \rho A\frac{\partial w}{\partial t}; \quad \frac{\partial F}{\partial w_x} = GAk\left(\phi - \frac{\partial w}{\partial x}\right) - p\frac{\partial w}{\partial x};$$

$$\frac{\partial F}{\partial w_{xx}} = -\frac{1}{4}GAl^2\left(\frac{\partial \phi}{\partial x} + \frac{\partial^2 w}{\partial x^2}\right) \tag{3.23}$$

Substituting Eqs. (3.23) into Eqs. (3.21) and (3.22) leads to the governing equations:

$$-\rho I\frac{\partial^2 \phi}{\partial t^2} + GAk\left(-\phi + \frac{\partial w}{\partial x}\right) + EI\frac{\partial^2 \phi}{\partial x^2} + \frac{GAl^2}{4}\left(\frac{\partial^2 \phi}{\partial x^2} + \frac{\partial^3 w}{\partial x^3}\right) = 0 \tag{3.24}$$

$$-\rho A\frac{\partial^2 w}{\partial t^2} + p\frac{\partial^2 w}{\partial x^2} + GAk\left(-\frac{\partial \phi}{\partial x} + \frac{\partial^2 w}{\partial x^2}\right) - \frac{GAl^2}{4}\left(\frac{\partial^3 \phi}{\partial x^3} + \frac{\partial^4 w}{\partial x^4}\right) = q \tag{3.25}$$

which hold on $x \in\leq (0, L)$. The integration by part of Eq. (3.20) yields the boundary conditions to be imposed at the ends of the beam ($x = 0$ and $x = L$):

- Either ϕ is specified or $- EI\frac{\partial \phi}{\partial x} - \frac{GAl^2}{4}\left(\frac{\partial \phi}{\partial x} + \frac{\partial^2 w}{\partial x^2}\right) = 0$ $\qquad$ (3.26)

- Either w is specified or $GAk\left(\phi - \frac{\partial w}{\partial x}\right) + \frac{GAl^2}{4}\left(\frac{\partial^2 \phi}{\partial x^2} + \frac{\partial^3 w}{\partial x^3}\right) - p\frac{\partial w}{\partial x} = 0$ $\quad$ (3.27)

- Either $\frac{\partial w}{\partial x}$ is specified or $- \frac{GAl^2}{4}\left(\frac{\partial \phi}{\partial x} + \frac{\partial^2 w}{\partial x^2}\right) = 0$ $\qquad$ (3.28)

Free vibration analysis

For a free vibration analysis, the transverse load q and the compressive load p will be eliminated from the equations to obtain:

$$-\rho I\frac{\partial^2 \phi}{\partial t^2} + GAk\left(-\phi + \frac{\partial w}{\partial x}\right) + EI\frac{\partial^2 \phi}{\partial x^2} + \frac{GAl^2}{4}\left(\frac{\partial^2 \phi}{\partial x^2} + \frac{\partial^3 w}{\partial x^3}\right) = 0 \tag{3.29}$$

$$-\rho A \frac{\partial^2 w}{\partial t^2} + G A k \left(-\frac{\partial \phi}{\partial x} + \frac{\partial^2 w}{\partial x^2} \right) - \frac{G A l^2}{4} \left(\frac{\partial^3 \phi}{\partial x^3} + \frac{\partial^4 w}{\partial x^4} \right) = 0 \qquad (3.30)$$

Buckling analysis

For a buckling analysis, the dependence of the equations on time is suppressed and the transverse load q along with the inertia forces are eliminated from the equations to obtain:

$$G A k \left(-\phi + \frac{dw}{dx} \right) + E I \frac{d^2 \phi}{dx^2} + \frac{G A l^2}{4} \left(\frac{d^2 \phi}{dx^2} + \frac{d^3 w}{dx^3} \right) = 0 \qquad (3.31)$$

$$p \frac{d^2 w}{dx^2} + G A k \left(-\frac{d\phi}{dx} + \frac{d^2 w}{dx^2} \right) - \frac{G A l^2}{4} \left(\frac{d^3 \phi}{dx^3} + \frac{d^4 w}{dx^4} \right) = 0 \qquad (3.32)$$

3.3 Finite Element Formulations

It was demonstrated in Chap. 2 that the four degrees of freedom finite element model gives reasonably good predictions for the static deflections of the microscale Timoshenko beam. Consequently, this same reduced-order model will be deployed here to evaluate the effect of the material length-scale parameter on the free vibration and buckling behaviours.

Free vibration analysis

We will discretize Eqs. (3.29) and (3.30) using the following representations of the field variables w and ϕ:

$$w(x, t) = \left[N_{wj}(x) \right] \{ w_j^e(t) \} = [N_w] \{ w_e(t) \}; \text{ where } \{ w_e \} = \begin{Bmatrix} w_1(t) \\ \phi_1(t) \\ w_2(t) \\ \phi_2(t) \end{Bmatrix} \qquad (3.33)$$

$$\phi(x, t) = \left[N_{\phi j}(x) \right] \{ w_j^e(t) \} = [N_\phi] \{ w_e(t) \} \qquad (3.34)$$

With the aid of the shape functions, N_w and N_ϕ, (defined in Chap. 2), the discretized time-dependent equations governing the free oscillatory motion of the microstructure-dependent Timoshenko beam takes the form:

$$M_{tim} \left(\frac{d^2 w_e(t)}{dt^2} \right) + K_{tim} w_e(t) = 0 \qquad (3.35)$$

where M_{tim} and K_{tim} denote mass and stiffness matrices, respectively. The explicit expressions for these matrices are:

$$K_{tim} = [K_{bending}] + [K_{shear}] + [K_{mcst}]$$

$$= \left(EI \int_0^{L_e} [B_\phi]^T [B_\phi] dx + kGA \int_0^{L_e} [B_\gamma]_4^T [B_\gamma]_4 dx \right.$$

$$\left. + (Gl^2 A) \int_0^{L_e} [B_x]_4^T [B_x]_4 dx \right) \tag{3.36}$$

$$M_{tim} = M_{tran} + M_{rot} \tag{3.37}$$

$$M_{tran} = \int_0^L \rho A N_w^T N_w dx \tag{3.38}$$

$$M_{rot} = \int_0^L \rho I N_\phi^T N_\phi dx \tag{3.39}$$

$$M_{tran} = \frac{\rho AL}{420(1+\varphi)^2}$$

$$\begin{bmatrix} 2(78 + 147\varphi + 70\varphi^2) & \frac{1}{2}L(44 + 77\varphi + 35\varphi^2) & 2(27 + 63\varphi + 35\varphi^2) & -\frac{1}{2}L(26 + 63\varphi + 35\varphi^2) \\ \frac{1}{2}L(44 + 77\varphi + 35\varphi^2) & \frac{1}{2}L^2(8 + 14\varphi + 7\varphi^2) & \frac{1}{2}L(26 + 63\varphi + 35\varphi^2) & -\frac{1}{2}L^2(6 + 14\varphi + 7\varphi^2) \\ 2(27 + 63\varphi + 35\varphi^2) & \frac{1}{2}L(26 + 63\varphi + 35\varphi^2) & 2(78 + 147\varphi + 70\varphi^2) & -\frac{1}{2}L(44 + 77\varphi + 35\varphi^2) \\ -\frac{1}{2}L(26 + 63\varphi + 35\varphi^2) & -\frac{1}{2}L^2(6 + 14\varphi + 7\varphi^2) & -\frac{1}{2}L(44 + 77\varphi + 35\varphi^2) & \frac{1}{2}L^2(8 + 14\varphi + 7\varphi^2) \end{bmatrix} \tag{3.40}$$

$$M_{rot} = \frac{\rho I}{30L(1+\varphi)^2} \begin{bmatrix} 36 & -3L(-1+5\varphi) & -36 & -3L(-1+5\varphi) \\ -3L(-1+5\varphi) & L^2(4+5\varphi+10\varphi^2) & 3L(-1+5\varphi) & L^2(-1-5\varphi+5\varphi^2) \\ -36 & 3L(-1+5\varphi) & 36 & 3L(-1+5\varphi) \\ -3L(-1+5\varphi) & L^2(-1-5\varphi+5\varphi^2) & 3L(-1+5\varphi) & L^2(4+5\varphi+10\varphi^2) \end{bmatrix} \tag{3.41}$$

With an assumption of a harmonic motion for the vibration problem, the vector of nodal displacements and accelerations will be represented by:

$$w_e(t) = w_e e^{i\omega t}; \ddot{w}_e(t) = -\omega^2 w_e e^{i\omega t} \tag{3.42}$$

An eigenvalue problem emerges if one substitutes Eq. (3.42) into (3.35). That is:

$$[-\omega^2 M_{tim} + K_{tim}] w_e e^{i\omega t} = 0 \tag{3.43}$$

The desired natural frequencies are then sought via a non-trivial solution search in the form:

$$|K - \omega^2 M| = 0 \tag{3.44}$$

Buckling Analysis

A buckling analyses demand that the discretization of Eqs. (3.31) and (3.32) are sought using the same representations of the field variables w and ϕ given in Eqs. (3.33) and (3.34). Doing this leads to another eigenvalue problem in the form:

$$\left[-\lambda K_g + K_{tim}\right]w_e = 0 \tag{3.45}$$

where K_{tim} is the same as defined in Eq. (3.36) and K_g is the stability matrix (which is a function of the compressive axial load p). Further, λ denotes the critical buckling load that has to be determined via a non-trivial solution search of:

$$\left|-\lambda K_g + K_{tim}\right| \tag{3.46}$$

The full form of the stability matrix is:

$$K_g = \int_0^L \left[\frac{dN_w}{dx}\right]^T p \left[\frac{dN_w}{dx}\right]dx = \frac{p}{10L(1+\varphi)^2}$$

$$\begin{bmatrix} 2\left(6+10\varphi+5\varphi^2\right) & L & -2\left(6+10\varphi+5\varphi^2\right) & L \\ L & \frac{1}{6}L^2\left(8+10\varphi+5\varphi^2\right) & -L & -\frac{1}{6}L^2\left(2+10\varphi+5\varphi^2\right) \\ -2\left(6+10\varphi+5\varphi^2\right) & -L & 2\left(6+10\varphi+5\varphi^2\right) & -L \\ L & -\frac{1}{6}L^2\left(2+10\varphi+5\varphi^2\right) & -L & \frac{1}{6}L^2\left(8+10\varphi+5\varphi^2\right) \end{bmatrix} \tag{3.47}$$

Section 3.4 details the implementation of the derived finite element model in the **R** programming language.

3.4   R Functions for Free Vibration and Buckling Analyses

This section lists a set of functions created to handle the free vibration and buckling analyses of microstructured-dependent microscale beams. The function for computing the stiffness matrix was presented in Chap. 2, and it is not repeated here for brevity sake. Nonetheless, all the functions presented here have been included in **microfiniteR**, hence it is encouraged that the package is loaded at the beginning of analyses to make the functions available.

Forming the mass matrix

```r
FormMassMTB <- function(youngmod, shearmod,
                        momentinertia, area,
                        shearfactor, poissonratio,
                        length, rho){

L <- length
ele_DOF <- 4
phi <- (12 * youngmod * momentinertia) / (shearfactor * area* shearmod * L^2)
beta3 <- (rho * area * length) / (420*(1 + phi)^2)
beta4 <- (rho * momentinertia)/(30*L*(1 + phi)^2)

m1 <- (156 + 294*phi + 140*phi^2)
m2 <- (22 + 38.5*phi + 17.5*phi^2)
m3 <- (54 + 126*phi + 70*phi^2)
m4 <- (13 + 31.5*phi + 17.5*phi^2)
m5 <- (4 + 7*phi + 3.5*phi^2)
m6 <- (3 + 7*phi + 3.5*phi^2)
m7 <- 37
m8 <- 3 - 15*phi
m9 <- 4 + 5*phi + 10*phi^2
m10 <- 1 + 5*phi - 5*phi^2

row1 <- c(m1, L*m2, m3, -L*m4)
row2 <- c(L*m2, L^2*m5, L*m4, -L^2*m6)
row3 <- c(m3, L*m4, m1, -L*m2)
row4 <- c(-L*m4,  L^2*m6, -L*m2, L^2*m5)

row21 <- c(m7, L*m8, -m7, L*m8)
row22 <- c(L*m8, L^2*m9, -L*m8, -L^2*m10)
row23 <- c(-m7, -L*m8, m7, -L*m8)
row24 <- c(L*m8, -L^2*m10, -L*m8, L^2*m9)

tran_massmatrix <- beta3 * matrix(c(row1,row2,row3,row4),
                                  nrow = ele_DOF,byrow = T)

rota_massmatrix <- beta4 * matrix(c(row21,row22,row23,row24),
                                  nrow = ele_DOF,byrow = T)

totalmass <- tran_massmatrix+rota_massmatrix
return(totalmass)

}
```

Forming the stability matrix

```r
FormStabilityMTB <- function(youngmod, shearmod,
                             momentinertia, area,
                             shearfactor, poissonratio, length){

L <- length
ele_DOF <- 4
phi <- (12 * youngmod * momentinertia) / (shearfactor * area* shearmod * L^2)
gamma <- 1/ (10*L*(1 + phi)^2)

s1 <- 2*(6 + 10*phi + 5*phi^2)
s2 <- L
s3 <- (1/6)*(8 + 10*phi + 5*phi^2)
s4 <- (1/6)*(2 + 10*phi + 5*phi^2)

row1 <- c(s1, s2, -s1, s2)
row2 <- c(s2, (L^2)*s3, -s2, -(L^2)*s4)
row3 <- c(-s1, -s2, s1, -s2)
row4 <- c(s2, -s4*(L^2), -s2, s3*(L^2))

stabilitymatrix <- gamma * matrix(c(row1,row2,row3,row4),
                                  nrow = ele_DOF, byrow = T)
return(stabilitymatrix)

}
```

Forming the global matrix

```
FormGlobalKMS=function(enumber, youngmod, shearmod,
                       momentinertia, area, shearfactor,
                       poissonratio, totallength, lengthscale,
                       rho, case){
  dof <- 4;
  dofpernode <- 2
  tdof <- (enumber + 1)*dofpernode;
  klist <- list();
  mlist <- list();
  slist <- list();
  length <- totallength/enum

# Stiffness matrix

  if(case == 1){

    for(j in seq(enumber)){

      klist[[j]]=FormStiffnessMTB(youngmod, shearmod,
                                  momentinertia, area,
                                  shearfactor, poissonratio,
                                  length, lengthscale);
    }

    globalmatrix <- matrix(0, tdof, tdof);
    Klist <- list();

    for(j in seq(enumber)){

      Klist[[j]] <- ExpandStiffnessMTB(tdof, klist[[j]], j, j+1)
      globalmatrix <- globalmatrix + Klist[[j]]
    }


  }

  # Mass matrix
  if(case == 2){

    for(j in seq(enumber)){

      mlist[[j]]=FormMassMTB(youngmod, shearmod,
                             momentinertia, area,
                             shearfactor, poissonratio,
                             length, rho);
    }

    globalmatrix <- matrix(0,tdof,tdof);
    Mlist <- list();

    for(j in seq(enumber)){

      Mlist[[j]] <- ExpandStiffnessMTB(tdof, mlist[[j]], j, j+1)
      globalmatrix <- globalmatrix + Mlist[[j]]
    }
  }

  # Stability matrix
  if(case == 3){

    for(j in seq(enumber)){

      slist[[j]]=FormStabilityMTB(youngmod, shearmod,
                                  momentinertia, area,
                                  shearfactor, poissonratio,
                                  length);
    }

    globalmatrix <- matrix(0, tdof, tdof);
    Slist <- list();

    for(j in seq(enumber)){

      Slist[[j]] <- ExpandStiffnessMTB(tdof, slist[[j]], j, j+1)
      globalmatrix <- globalmatrix + Slist[[j]]
    }
  }

  return(globalmatrix)

}
```

Finding the natural frequencies (radian/seconds)

```
FindFrequenciesRad <- function(reducedM, reducedK)
{
  massinv <- solve(reducedM);
  productMK <- massinv%*%reducedK
  syseigen <- eigen(productMK)
  sortedfreq <- sort(syseigen$values)

  return(sqrt(sortedfreq))
}
```

Finding the critical buckling loads

```
FindCriticalLoad <- function(reducedM, reducedK)
{
  stabilityinv <- solve(reducedS)
  productSK <- stabilityinv %*% reducedK
  criticalloads <- eigen(productSK)
  sortedloads <- sort(criticalloads$values)[1]
  return(sortedloads)
}
```

3.5 Free Vibration and Buckling Problems with the Implemented R Functions

Example 3.1 Natural frequencies of a simply-supported microstructure-dependent beam

Problem Compute the first three natural frequencies of a simply-supported microscale Timoshenko beam for values[2] of l/h from 0 to 1 at an interval of 0.2. Discretize the beam into ten elements and take: $E = 1.44$ GPa; $v = 0.38$; $k = \frac{5+5v}{6+5v}$; $h = 17.6\ \mu\text{m}$; $b = 20\ h$; $L = 20\ h$.

Solution

Figure 3.1 shows the microscale beam that has been discretized into ten elements. Following the discretization, each element has a length $L_e = 20\ h/10$. The nodes are numbered from left to right, with node 1 and node 11 located at the leftmost and rightmost ends, respectively.

Step 1—Load the **microfiniteR** package, supply the material/geometric properties of the structure and declare the number of elements.

---

[2]As is commonly used in the literature on this topic, this value represents the ratio of material length-scale parameter to the thickness of the beam.

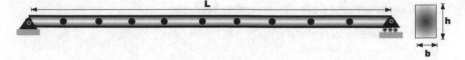

Fig. 3.1 A 10-element simply-supported microscale beam

```
library(microfiniteR)
#Material and geometric properties
E = 1.44e6
den = 1220
pratio = 0.38
G = E/(2*(1 + pratio))
sf = (5+5*pratio)/(6+5*pratio)
h = 17.6e-3
b = 2*h
momenti = (b*h^3)/12;
area = b*h
totallen = 20*h
alpha = 0.2          #ratio of lengthscale to thickness
lscale = alpha * h


enum = 10            #number of elements
elen = 20*h/enum
nnum = enum+1        #number of nodes
tdof = nnum*2        #total degrees of freedom
```

Step 2—Form the global stiffness and mass matrices using the function
FormGlobalKMS().

```
bigK=FormGlobalKMS(enum, E, G, momenti, area, sf, pratio, totallen, lscale, den, 1);
bigM=FormGlobalKMS(enum, E, G, momenti, area, sf, pratio, totallen, lscale, den, 2);
```

The output of the above produces the global stiffness and mass matrices, respectively. The output has been suppressed and not displayed for brevity sake. Having formed the global matrices, it is now possible to obtain the reduced matrices by applying the boundary conditions.

Step 3—Apply the boundary condition(s) on the global stiffness and mass matrices to obtain the corresponding reduced matrices.

For the boundary conditions, nodes 1 and 11 are hinged, and we therefore call the function **HingeNodes()** on these nodes to eliminate the vertical displacements. Further, the output of **HingeNodes()** is used with **ExtractFreeRows()** to obtain the pointers to the degrees of freedom of the unrestrained nodes. Finally, the output of **ExtractFreeRows()** is used with **FormReducedStiffness()** to establish the reduced the stiffness matrices.

Table 3.1 Natural frequencies of simply-supported microstructure-dependent beams

l/h	λ_1		λ_2		λ_3	
	Current	Reddy [20]	Current	Reddy [20]	Current	Reddy [20]
0.2	10.643	10.65	42.003	42.06	92.621	92.78
0.4	12.777	12.80	50.271	50.52	110.351	110.34
0.6	15.699	15.73	61.582	61.01	134.599	136.39
0.8	19.048	19.08	74.553	75.05	162.425	164.51
1.0	22.636	22.66	88.457	88.84	192.288	193.82

```
freenodes=ExtractFreeRows(tdof, HingeNodes(c(1,nnum)));
reducedK=FormReducedMatrix(bigK, freenodes);
reducedM=FormReducedMatrix(bigM, freenodes);
```

Step 4—Find the natural frequencies.

```
naturalfreq = FindFrequenciesRad(reducedM,reducedK)
firstthree = cbind(naturalfreq[1:3]);
nondim_freq = sqrt((firstthree^2) * (den * area * totallen^4)/(E * momenti));
nondim_freq
```

The first line of the preceding code snippet returns the natural frequencies in radians per seconds, and in the second line we extract the first three natural frequencies. In the third line, we convert the frequencies to a non-dimensional form $\left(\lambda = \frac{\rho A L^4 \omega^2}{EI}\right)$ so as to allow comparison with published studies, specifically with Reddy [20]. The output of the last line therefore yields:

```
> nondim_freq
          [,1]
[1,] 10.64291
[2,] 42.00260
[3,] 92.62078
```

Additional results are listed in Table 3.1.

Example 3.2 Critical loads of a simply-supported microstructure-dependent beam

Problem Compute the critical loads of a simply-supported microscale Timoshenko beam for values[3] of l/h that varies as $\frac{1}{10}, \frac{1}{9}, \frac{1}{7}, \frac{1}{5}, \frac{1}{3}, 1$. Take: $E = 1.44$ GPa; $v = 0.38; k = \frac{5+5v}{6+5v}; h = 17.6$ μm; $b = 20 h; L = 20 h$.

Solution
For this problem, we also consider a 10-element discretization, leading to each element having a length $L_e = 20 h/10$. The results are listed in Table 3.2. The combined

---

[3]These values have been chosen to allow comparison with published results.

Table 3.2 Buckling loads of a simply-supported microstructure-dependent beam

l/h	FEM (Present)	Analytical [23]
1/10	10.230	10.2987
1/9	10.329	10.3994
1/7	10.671	10.7453
1/5	11.500	11.5861
1/3	14.509	14.6375
1	51.732	52.7809

lines of code for the case when $l/h = 1/10$ are grouped together in the code snippet below. For the other cases, only the value of ratio of material lengthscale to thickness is changed accordingly.

```
library(microfiniteR)
#Material and geometric properties
E = 1.44e6
den = 1220
pratio = 0.38
G = E/(2*(1 + pratio))
sf = (5+5*pratio)/(6+5*pratio)
h = 17.6e-3
b = 2*h
momenti = (b*h^3)/12;
area = b*h
totallen = 20*h
alpha = 1/10.        #ratio of lengthscale to thickness
lscale = alpha * h


enum = 10        #number of elements
elen = 20*h/enum
nnum = enum+1    #number of nodes
tdof = nnum*2    #total degrees of freedom



bigK=FormGlobalKMS(enum, E, G, momenti, area, sf, pratio, totallen, lscale, den, 1);
bigS=FormGlobalKMS(enum, E, G, momenti, area, sf, pratio, totallen, lscale, den, 3);


freenodes=ExtractFreeRows(tdof, HingeNodes(c(1,nnum)));freenodes
reducedK=FormReducedMatrix(bigK, freenodes);
reducedS=FormReducedMatrix(bigS, freenodes);


criticalload = FindCriticalLoad(reducedM,reducedK)
nondim_criticalload = criticalload * totallen^2/(E * momenti);
nondim_criticalload
```

Example 3.3 Natural frequencies of clamped-sliding and clamped-clamped microstructure-dependent beams

Table 3.4 Natural frequencies of scale-dependent beams with clamped-sliding and clamped-clamped boundary conditions

Modes	Clamped-Sliding		Clamped-Clamped	
	Classical[a]	$l = h$	Classical	$l = h$
λ_1	5.569	12.734	21.991	48.403
λ_2	29.700	65.973	59.323	124.265
λ_3	71.857	153.510	113.161	225.092

[a]This refers to the case when $l = 0$

Problem Compute and compare the first three natural frequencies of microscale beams with the two different boundary conditions shown in Table 3.3.

$$E = 1.44\ \text{GPa};$$

$$G = \frac{E}{2(1 + v)};$$

$$v = 0.38; k = \frac{5 + 5v}{6 + 5v}; h = 17.6\,\mu\text{m};$$

$$b = 20\ h; I = \frac{bh^3}{12}; A = bh; L = 12\ h; l = h.$$

Solution

For each of the two specified boundary conditions, the variations of the first three free vibration modes with $l = h$ have been computed and listed in Table 3.4. For the purpose of comparison, the case of $l = 0$, which leads to the classical model is also computed and presented, and it is observed that the computed values for the classical case are in good agreement with the predictions in Blevins [24].[4]

The code snippet for the computation of the natural frequencies of the clamped-sliding beam is given next.

---

[4]In this text, $\lambda = \sqrt[2]{\rho A L^4 \omega^2 / EI}$. However, it is pointed out that the non-dimensional frequency values in *Formulas for dynamics, acoustics and vibration,* are found by taking the square root of the values presented here.

```
library(microfiniteR)
#Material and geometric properties
E = 1.44e6
den = 1220
pratio = 0.38
G = E/(2*(1 + pratio))
sf = (5+5*pratio)/(6+5*pratio)
h = 17.6e-3
b = 2*h
momenti = (b*h^3)/12;
area = b*h
totallen = 20*h
alpha = 1          #ratio of lengthscale to thickness. For classical alpha=0.
lscale = alpha * h


enum = 30          #number of elements
elen = 20*h/enum
nnum = enum+1      #number of nodes
tdof = nnum*2      #total degrees of freedom


bigK=FormGlobalKMS(enum, E, G, momenti, area, sf, pratio, totallen, lscale, den, 1);
bigM=FormGlobalKMS(enum, E, G, momenti, area, sf, pratio, totallen, lscale, den, 2);


freenodes=ExtractFreeRows(tdof, c(FixNodes(1),nnum*2));freenodes
reducedK=FormReducedMatrix(bigK, freenodes);
reducedM=FormReducedMatrix(bigM, freenodes);


naturalfreq = FindFrequenciesRad(reducedM,reducedK)
firstthree = cbind(naturalfreq[1:3]);
nondim_freq = sqrt((firstthree^2) * (den * area * totallen^4)/(E * momenti));
nondim_freq
```

The non-dimensional frequencies from the above is given next:

```
> nondim_freq
           [,1]
[1,]   12.73367
[2,]   65.97253
[3,]  153.51017
```

To compute the frequencies of the clamped-clamped beam, a change in the boundary conditions is made, and the resulting lines of code involving the required changes are presented next.

```
library(microfiniteR)
#Material and geometric properties
E = 1.44e6
den = 1220
pratio = 0.38
G = E/(2*(1 + pratio))
sf = (5+5*pratio)/(6+5*pratio)
h = 17.6e-3
b = 2*h
momenti = (b*h^3)/12;
area = b*h
totallen = 20*h
alpha = 1          #ratio of lengthscale to thickness
lscale = alpha * h


enum = 40          #number of elements
elen = 20*h/enum
nnum = enum+1      #number of nodes
tdof = nnum*2      #total degrees of freedom


bigK=FormGlobalKMS(enum, E, G, momenti, area, sf, pratio, totallen, lscale, den, 1);
bigM=FormGlobalKMS(enum, E, G, momenti, area, sf, pratio, totallen, lscale, den, 2);


freenodes=ExtractFreeRows(tdof, FixNodes(c(1, nnum)));
reducedK=FormReducedMatrix(bigK, freenodes);
reducedM=FormReducedMatrix(bigM, freenodes);


naturalfreq = FindFrequenciesRad(reducedM, reducedK)
firstthree = cbind(naturalfreq[1:3]);
nondim_freq = sqrt((firstthree^2) * (den * area * totallen^4)/(E * momenti));
nondim_freq
```

Running the above produces the following:

```
2.  > nondim_freq
3.          [,1]
4.  [1,]  48.40346
5.  [2,] 124.26475
6.  [3,] 225.09189
```

As mentioned earlier, if the value of the variable **alpha** in the code snippet is changed to zero, the results obtained match those presented by Blevins [24]. Minimal changes can be made to the code to handle other types of non-classical boundary conditions for buckling analyses.

3.6 Summary

In this chapter, the equations that govern the free vibration and buckling behaviours of microstructure-dependent microscale Timoshenko beams are derived with consideration of the size-dependent effect. The finite element discretized equivalent of the equations are then established. A further set of **R** functions are provided to facilitate computations of the natural frequencies and buckling loads of the Timoshenko microscale beam. The developed functions are included in **microfiniteR**. Examples are presented to demonstrate the validity of results obtained using the package. While only a few cases have been analyzed and presented, changes can be made to the material and geometric parameters as well as boundary conditions to allow further extensions of the codes to deal with some other cases.

References

1. S.S. Rao, *Vibration of Continuous Systems* (Wiley, Hoboken, 2007)
2. L.N. Virgin, *Vibration of Axially Loaded Structures* (Cambridge University Press, New York, NY, 2007)
3. M.A. Wahab, *Dynamics and Vibration: An Introduction* (Wiley, Hoboken, 2008)
4. J.N. Reddy, *Introduction to the Finite Element Method* (McGraw-Hill, New York, 1993)
5. M. Petyt, *Introduction to Finite Element Vibration Analysis* (Cambridge University Press, Cambridge, 1998)
6. M.A. Bhatti, *Advanced Topics in Finite Element Analysis of Structures: With Mathematica and MATLAB Computations* (Wiley, Hoboken, 2006)
7. J. Reddy, On the dynamic behaviour of the Timoshenko beam finite elements. Sadhana **24**, 175–198 (1999)
8. S. Saadon, O. Sidek, A review of vibration-based MEMS piezoelectric energy harvesters. Energy Convers. Manag. **52**, 500–504 (2011)
9. M.I. Younis, *MEMS Linear and Nonlinear Statics and Dynamics* (Springer, US, 2011)
10. W.K. Schomburg, *Introduction to Microsystem Design* (Springer, Heidelberg, 2011)
11. I.J. Phelps, *Mechanical Characterization of MEMS Bi-stable Buckled Diaphragms* (2013)
12. S. Kong, S. Zhou, Z. Nie, K. Wang, The size-dependent natural frequency of Bernoulli-Euler micro-beams. Int. J. Eng. Sci. **46**, 427–437 (2008)
13. H.M. Ma, X.L. Gao, J.N. Reddy, A microstructure-dependent Timoshenko beam model based on a modified couple stress theory. J. Mech. Phys. Solids **56**, 3379–3391 (2008)
14. K. Mustapha, Size-dependent flexural dynamics of ribs-connected polymeric micropanels. CMC: Comput. Mater. Continua **42**, 141–174 (2014)
15. K. Mustapha, Z. Zhong, Wave propagation characteristics of a twisted micro scale beam. Int. J. Eng. Sci. **53**, 46–57 (2012)
16. K.B. Mustapha, Modeling of a functionally graded micro-ring segment for the analysis of coupled extensional–flexural waves. Compos. Struct. **117**, 274–287 (2014)
17. A. Nateghi, M. Salamat-talab, J. Rezapour, B. Daneshian, Size dependent buckling analysis of functionally graded micro beams based on modified couple stress theory. Appl. Math. Model. **36**, 4971–4987 (2012)
18. J.N. Reddy, *Energy Principles and Variational Methods in Applied Mechanics*, 2nd edn. (Wiley, Hoboken, NJ, 2002)
19. F. Yang, A.C.M. Chong, D.C.C. Lam, P. Tong, Couple stress based strain gradient theory for elasticity. Int. J. Solids Struct. **39**, 2731–2743 (2002)

20. J.N. Reddy, Microstructure-dependent couple stress theories of functionally graded beams. J. Mech. Phys. Solids **59**, 2382–2399 (2011)
21. E.B. Magrab, *Vibrations of Elastic Systems: With Applications to MEMS and NEMS*, vol 184 (Springer Science & Business Media, 2012)
22. J.N. Reddy, *Energy Principles and Variational Methods in Applied Mechanics* (Wiley, Hoboken, 2002)
23. B. Akgöz, Ö. Civalek, Strain gradient elasticity and modified couple stress models for buckling analysis of axially loaded micro-scaled beams. Int. J. Eng. Sci. **49**, 1268–1280 (2011)
24. R.D. Blevins, *Formulas for Dynamics, Acoustics and Vibration* (Wiley, Hoboken, 2015)

Chapter 4
Bending and Vibration of Microstructure-Dependent Kirchhoff Microplates and Finite Element Implementations with R

Abstract In Chaps. 2 and 3, our concern was on structures for which only one of the geometric dimensions dominates. This resulted in the approximations of such three-dimensional (3D) structures as one-dimensional beams. However, for 3D structures having two dominating planar geometric dimensions, a more appropriate approximation demands the move from the use of beam theories to advance structural models. This chapter deals with the most elementary of such theories, the so-called classical plate theory following the Kirchhoff's hypothesis. Our interest being on the microstructured-dependent behaviour, the theoretical treatment is again approached from the perspectives of the modified couple stress theory to establish the appropriate size-dependent model of a microscale plate for static and free vibration analyses. Attention is given to the implementation of the finite element solutions for these analyses in the R programming language. Although the developed finite element model is a non-conforming rectangular element, it provides a straightforward way to demonstrate the influence of material length-scale parameters on the bending and dynamic behaviour of plates with various boundary conditions.

4.1 Introduction

As narrated in a number of excellent texts on the theory of plates, many notable historical figures from Sophie Germain, Lagrange, Lévy, Navier, Poisson and Kirchhoff contributed to our fundamental understanding of the theoretical underpinning of mechanics of plates [1–4]. The significance of the topic is not unconnected with the important roles played by plates in several applications. This ranges from transport infrastructures (marine, aerospace and automotive), acoustic radiators to building, and machinery. In the domain of microsystems, microscale plate-like structures function as one of the primary elements in capacitive transducers/parallel-plate electrostatic actuators, radio frequency micro-electro-mechanical system filters, diaphragm-based pressure sensing, torsional and extensional resonators etc. [5–8]. In some of these sensitive devices and resonators, the static deflection and internal resistance loads are of utmost

© The Author(s), under exclusive license to Springer Nature Singapore Pte Ltd. 2019
K. B. Mustapha, *R for Finite Element Analyses of Size-dependent Microscale Structures*, SpringerBriefs in Computational Mechanics,
https://doi.org/10.1007/978-981-13-7014-4_4

interest, and in others, the quality factor closely relates to the frequency response [9]. An accurate mathematical modelling is therefore required for design considerations.

Some of the earliest studies on microplates (for use as filters, distributed sensors for chemical and mechanical measurement, and detectors etc.) relate to the examination of the effect of van der Waals force and prediction of electrostatic deflections and pull-in instabilities [10–13]. In recent years, several researchers have investigated the effect of material length-scale on the static and dynamic behaviours of microscale plates. In this chapter, the primary aim is not so much to propose a new model but to re-examine an earlier model of microplates presented by Tsiatas [14] and Jomehzadeh et al. [15]. A clearly incomplete list of studies that align closely with the current line of presentation includes [15–22], all of which offer formulations that are consistent with the theme of the previous chapters (i.e. the modified couple stress theory). Nevertheless, some studies, such as those by Lazopoulos [23, 24], have approached the issue from the perspective of the strain gradient theory. Meanwhile, a good number of studies on size-dependent models of microplates are still very much oriented towards analytical solutions. It is well-known that analytical solutions to plate vibration problems faces a considerable obstacle when dealing with complicated boundary conditions. Therefore this chapter examines a finite element model for the static bending and vibration response analyses of microstructure-dependent microscale plates. In Sect. 4.2, the distributed mathematical model of microscale plates with a size-dependent property is reviewed. This is then followed by the finite element formulation of the model in Sect. 4.3. The collection of $\mathbf{R}$ functions implemented to support the numerical evaluation of the model is located in Sect. 4.4, and this is followed by a summary of the chapter.

4.2 Governing Differential Equations

We refer the microplate to the system of rectangular coordinates xyz, such that x and y lie in the middle plane and the faces of the microplate are positioned at $z = \pm h/2$. The components of the displacements, following the Kirchhoff's plate hypothesis, takes the form [25, 26]:

$$u_1(x, y, z, t) = -z\frac{\partial w(x, y, t)}{\partial x}; u_2(x, y, z, t) = -z\frac{\partial w(x, y, t)}{\partial y}; u_3 = w(x, y, t)$$

(4.1)

where w denotes the transverse displacement of the neutral plane. Using Eq. (4.1), one will be able to obtain the non-zero components of the strain tensor, the rotation vector, and the symmetric curvature tensor, as follows.[1]

---

[1]The displacement-strain, stress-strain and the symmetric curvature-couple stress relations are stated in Chap. 2.

Non-zero components (strain tensor)

$$\varepsilon_{xx} = -z\frac{\partial^2 w(x, y, t)}{\partial x^2}; \quad \varepsilon_{yy} = -z\frac{\partial^2 w(x, y, t)}{\partial y^2} \tag{4.2}$$

$$\varepsilon_{xy} = \varepsilon_{yx} = -z\frac{\partial^2 w(x, y, t)}{\partial x \partial y} \tag{4.3}$$

Non-zero components (rotation vector)

$$\theta_x = \frac{1}{2}\left(\frac{\partial u_3}{\partial z} - \frac{\partial u_2}{\partial y}\right) = \frac{\partial w(x, y, t)}{\partial y}; \theta_y = \frac{1}{2}\left(\frac{\partial u_1}{\partial z} - \frac{\partial u_3}{\partial x}\right) = -\frac{\partial w(x, y, t)}{\partial x} \tag{4.4}$$

Non-zero components (curvature tensor)

$$\chi_{xx} = \frac{\partial \theta_x}{\partial x} = \frac{\partial^2 w(x, y, t)}{\partial y \partial x}; \chi_{yy} = \frac{\partial \theta_y}{\partial y} = -\frac{\partial^2 w(x, y, t)}{\partial x \partial y}; \tag{4.5}$$

$$\chi_{xy} = \frac{1}{2}\left(\frac{\partial \theta_x}{\partial y} + \frac{\partial \theta_y}{\partial x}\right) = \frac{1}{2}\left(\frac{\partial^2 w(x, y, t)}{\partial y^2} - \frac{\partial^2 w(x, y, t)}{\partial x^2}\right) \tag{4.6}$$

Non-zero components (stress and couple stress tensors)

$$\sigma_{xx} = \frac{E}{1 - v^2}\left(\varepsilon_{xx} + v\varepsilon_{yy}\right); \sigma_{yy} = \frac{E}{1 - v^2}\left(\varepsilon_{yy} + v\varepsilon_{xx}\right); \sigma_{xy} = G\varepsilon_{xy} = \frac{E}{1 + v}\varepsilon_{xy} \tag{4.7}$$

$$m_{xx} = 2Gl^2\chi_{xx} = \frac{El^2}{1 + v}\chi_{xx}; m_{yy} = \frac{El^2}{1 + v}\chi_{yy}; m_{xy} = \frac{El^2}{1 + v}\chi_{xy} \tag{4.8}$$

The size-dependent strain energy (Π_s) that arises from the deformation of the microplate is a superposition of the classical strain energy term (Π_c) and the one that arises from the modified couple stress theory (Π_{mcst}), and it takes the form:

$$\Pi_s = \Pi_c + \Pi_{mcst}$$

$$= \frac{1}{2}\iint\limits_{R}\int\limits_{-\frac{h}{2}}^{\frac{h}{2}}\left(\sigma_{xx}\varepsilon_{xx} + \sigma_{yy}\varepsilon_{yy} + 2\sigma_{xy}\varepsilon_{xy}\right)dz dA$$

$$+ \frac{1}{2}\iint\limits_{R}\int\limits_{-\frac{h}{2}}^{\frac{h}{2}}\left(m_{xx}\chi_{xx} + m_{yy}\chi_{yy} + 2m_{xy}\chi_{xy}\right)dz dA \tag{4.9}$$

By making use of Eqs. (4.2) through Eq. (4.8) and integrating over the thickness, one may re-express Eq. (4.9) as:

$$\Pi_s = \frac{Eh^3}{24(1 - v^2)}\iint\limits_{R}\left[\begin{array}{l}\left(\frac{\partial^2 w}{\partial y^2}\right)^2 + \left(\frac{\partial^2 w}{\partial x^2}\right)^2 + 2(1 - v)\left(\frac{\partial^2 w(x, y, t)}{\partial x \partial y}\right)^2 \\ + 2v\left(\frac{\partial^2 w}{\partial y^2}\right)\left(\frac{\partial^2 w}{\partial x^2}\right)\end{array}\right]dx dy$$

$$
+ \frac{Ehl^2}{2(1+v)} \iint_R \left[\begin{array}{c} \left(\frac{\partial^2 w}{\partial y^2}\right)^2 + \left(\frac{\partial^2 w}{\partial x^2}\right)^2 + 2\left(\frac{\partial^2 w(x,y,t)}{\partial x \partial y}\right)^2 \\ - 2\left(\frac{\partial^2 w}{\partial y^2}\right)\left(\frac{\partial^2 w}{\partial x^2}\right) \end{array} \right] dxdy \qquad (4.10)
$$

The total work done by the distributed load $q(x, y)$ acting on the mid-plane surface of the microplate takes the form:

$$
\Pi_W = - \iint_R q(x,y)w(x,y)dxdy \qquad (4.11)
$$

Putting Eqs. (4.10) and (4.11) together yields the total potential energy (Π_P) as[2]:

$$
\begin{aligned}
\Pi_P &= \Pi_s + \Pi_W \\
&= \frac{D}{2} \iint_R \left[\begin{array}{c} \left(\frac{\partial^2 w}{\partial y^2}\right)^2 + \left(\frac{\partial^2 w}{\partial x^2}\right)^2 + 2(1-v)\left(\frac{\partial^2 w(x,y,t)}{\partial x \partial y}\right)^2 \\ + 2v\left(\frac{\partial^2 w}{\partial y^2}\right)\left(\frac{\partial^2 w}{\partial x^2}\right) \end{array} \right] dxdy \\
&\quad + \frac{Ehl^2}{2(1+v)} \iint_R \left[\begin{array}{c} \left(\frac{\partial^2 w}{\partial y^2}\right)^2 + \left(\frac{\partial^2 w}{\partial x^2}\right)^2 + 2\left(\frac{\partial^2 w(x,y,t)}{\partial x \partial y}\right)^2 \\ - 2\left(\frac{\partial^2 w}{\partial y^2}\right)\left(\frac{\partial^2 w}{\partial x^2}\right) \end{array} \right] dxdy \\
&\quad - \iint_R q(x,y)w(x,y)dxdy \qquad (4.12)
\end{aligned}
$$

where $D = \frac{Eh^3}{12(1-v^2)}$. The kinetic energy is as follows:

$$
\begin{aligned}
\Pi_K &= \frac{1}{2} \iint_R \int_{-\frac{h}{2}}^{\frac{h}{2}} \rho \left[\left(\frac{\partial u_1}{\partial t}\right)^2 + \left(\frac{\partial u_2}{\partial t}\right)^2 + \left(\frac{\partial u_3}{\partial t}\right)^2 \right] dzdxdy \\
&= \frac{1}{2} \iint_R \left[I_0\left(\frac{\partial w}{\partial t}\right)^2 + I_1\left(\frac{\partial^2 w}{\partial y \partial t}\right)^2 + I_1\left(\frac{\partial^2 w}{\partial x \partial t}\right)^2 \right] dxdy \qquad (4.13)
\end{aligned}
$$

where $I_0 = \rho h; I_1 = \rho h^3/12$.

---

[2] Assuming isotropic material properties.

4.2.1 The Extended Hamilton's Principle

The Hamilton's principle (as stated in Chap. 3) demands the stationarity of the integral of the energy functional (I) in the configuration space between the time interval t_1 and t_2, where I is as defined in Eq. (4.14) [27]:

$$I = \int_{t_1}^{t_2} \mathcal{L}\, dt; \quad \mathcal{L} = \Pi_K - \Pi_P \tag{4.14}$$

By combining Eqs. (4.12) and (4.13), we can re-write Eq. (4.14) as:

$$
\begin{aligned}
l = \int_{t_1}^{t_2} \Big\{ & \frac{1}{2} \int \int_R \left[I_0 \left(\frac{\partial w}{\partial t} \right)^2 + I_1 \left(\frac{\partial^2 w}{\partial y \partial t} \right)^2 + I_1 \left(\frac{\partial^2 w}{\partial x \partial t} \right)^2 \right] dxdy \\
& - \frac{D}{2} \int \int_R \left[\left(\frac{\partial^2 w}{\partial y^2} \right)^2 + \left(\frac{\partial^2 w}{\partial x^2} \right)^2 + 2(1-v) \left(\frac{\partial^2 w(x, y, t)}{\partial x \partial y} \right)^2 + 2v \left(\frac{\partial^2 w}{\partial y^2} \right) \left(\frac{\partial^2 w}{\partial x^2} \right) \right] dxdy \\
& - \frac{Ehl^2}{2(1+v)} \int \int_R \left[\left(\frac{\partial^2 w}{\partial y^2} \right)^2 + \left(\frac{\partial^2 w}{\partial x^2} \right)^2 + 2 \left(\frac{\partial^2 w(x, y, t)}{\partial x \partial y} \right)^2 - 2 \left(\frac{\partial^2 w}{\partial y^2} \right) \left(\frac{\partial^2 w}{\partial x^2} \right) \right] dxdy \\
& - \int \int_R q(x, y) w(x, y) dxdy \Big\} dt
\end{aligned}
\tag{4.15}
$$

which, in terms of a functional F, can be re-cast as[3]:

$$I = \int_{t_2}^{t_2} \left\{ \iint_R F\left(x, y, t, w, w_t, w_{yt}, w_{xt}, w_{yy}, w_{xx}\right) dxdy \right\} dt \tag{4.16}$$

Equation (4.16) is a variational problem with one temporal and two spatial dimensions. It has x, y and t as the independent variables and w as the dependent variable.

As done in Chap. 3, the variational derivative of $I(w)$ may be achieved by defining a family of trial functions:

$$\bar{w}(x, y, t) = w(x, y, t) + \epsilon \alpha(x, y, t) \tag{4.17}$$

Connecting Eqs. (4.17) with (4.16) leads to:

$$\bar{I}(\epsilon) = \int_{t_1}^{t_2} \left\{ \iint_R F\left(x, y, t, \bar{w}, \bar{w}_t, \bar{w}_{yt}, \bar{w}_{xt}, \bar{w}_{yy}, \bar{w}_{xx}, \bar{w}_{xy}\right) dxdy \right\} dt \tag{4.18}$$

---

[3]The subscript notations have been used to denote differential operators such that: $w_t = \frac{\partial w}{\partial t}$; $w_{xx} = \frac{\partial^2 w}{\partial x^2}$; $w_{yy} = \frac{\partial^2 w}{\partial y^2}$; $w_{xy} = \frac{\partial^2 w}{\partial x \partial x}$; $w_{xt} = \frac{\partial^2 w}{\partial x \partial t}$; $w_{yt} = \frac{\partial^2 w}{\partial y \partial t}$

The extremum of $\bar{I}(\epsilon)$ is obtained by differentiating Eq. (4.18). That is:

$$\left.\frac{\partial \bar{I}}{\partial \epsilon}\right|_{\epsilon=0} = 0 \tag{4.19}$$

Mapping the partial differential operator to $\bar{I}$ produces:

$$\frac{\partial \bar{I}}{\partial \epsilon} = \int_{t_1}^{t_2} \left\{ \iint_R \left(\frac{\partial F}{\partial \bar{w}} \frac{\partial \bar{w}}{\partial \epsilon} + \frac{\partial F}{\partial \bar{w}_t} \frac{\partial \bar{w}_t}{\partial \epsilon} + \frac{\partial F}{\partial \bar{w}_{tx}} \frac{\partial \bar{w}_{tx}}{\partial \epsilon} \right. \right.$$
$$\left. \left. + \frac{\partial F}{\partial \bar{w}_{ty}} \frac{\partial \bar{w}_{ty}}{\partial \epsilon} + \frac{\partial F}{\partial \bar{w}_{xx}} \frac{\partial \bar{w}_{xx}}{\partial \epsilon} + \frac{\partial F}{\partial \bar{w}_{yy}} \frac{\partial \bar{w}_{yy}}{\partial \epsilon} + \frac{\partial F}{\partial \bar{w}_{xy}} \frac{\partial \bar{w}_{xy}}{\partial \epsilon} dxdydt \right. \right. \tag{4.20}$$

Bearing in mind Eq. (4.17), we re-write Eq. (4.20) as:

$$\left.\frac{d\bar{I}}{d\epsilon}\right|_{\epsilon=0} = \int_{t_1}^{t_2} \left\{ \iint_R \left(\frac{\partial F}{\partial w}\alpha + \frac{\partial F}{\partial w_t}\alpha_t + \frac{\partial F}{\partial w_{tx}}\alpha_{tx} + \frac{\partial F}{\partial w_{ty}}\alpha_{ty} \right. \right.$$
$$\left. \left. + \frac{\partial F}{\partial w_{xx}}\alpha_{xx} + \frac{\partial F}{\partial w_{yy}}\alpha_{yy} + \frac{\partial F}{\partial w_{xy}}\alpha_{xy} \right) dxdy \right\} dt \tag{4.21}$$

Simplifying further produces:

$$\left.\frac{d\bar{I}}{d\epsilon}\right|_{\epsilon=0} = \int_{t_1}^{t_2} \left\{ \iint_R \left(\left(\frac{\partial F}{\partial w} - \frac{\partial}{\partial t}\frac{\partial F}{\partial w_t} \right)\alpha - \left(\frac{\partial}{\partial t}\frac{\partial F}{\partial w_{tx}}\alpha_x + \frac{\partial}{\partial t}\frac{\partial F}{\partial w_{ty}}\alpha_y \right) \right. \right.$$
$$\left. \left. + \frac{\partial F}{\partial w_{xx}}\alpha_{xx} + \frac{\partial F}{\partial w_{yy}}\alpha_{yy} + \frac{\partial F}{\partial w_{xy}} \right) \alpha_{xy} dxdy \right\} dt \tag{4.22}$$

The Green's theorem[4] is used simplify Eq. (4.22) to yield:

$$\int_{t_1}^{t_2} \left\{ \iint_R \left[\frac{\partial F}{\partial w} - \frac{\partial}{\partial t}\frac{\partial F}{\partial w_t} + \frac{\partial^2}{\partial t\partial x}\frac{\partial F}{\partial w_{tx}} + \frac{\partial^2}{\partial t\partial y}\frac{\partial F}{\partial w_{ty}} \right. \right.$$
$$\left. \left. + \frac{\partial^2}{\partial x^2}\frac{\partial F}{\partial w_{xx}} + \frac{\partial^2}{\partial y^2}\frac{\partial F}{\partial w_{yy}} + \frac{\partial^2}{\partial x\partial y}\frac{\partial F}{\partial w_{xy}} \right] \alpha \right\}$$
$$dxdydt + Series of Line Integrals \tag{4.23}$$

The decomposition of the series of line integrals in Eq. (4.23) will yield the natural boundary conditions, but it is not pursued further here for brevity sake (exhaustive discussions on this can be found in [25, 26, 28, 29]). Nevertheless, from Eq. (4.23), the Euler-Lagrange equation is obtained in the form of the following partial differential equation:

---

[4] $\iint \left[\frac{\partial P}{\partial x} + \frac{\partial Q}{\partial y} \right] dxdy = \oint (Pdy - Qdx)$

$$\frac{\partial F}{\partial w} - \frac{\partial}{\partial t}\frac{\partial F}{\partial w_t} + \frac{\partial^2}{\partial t \partial x}\frac{\partial F}{\partial w_{tx}} + \frac{\partial^2}{\partial t \partial y}\frac{\partial F}{\partial w_{ty}}$$

$$+ \frac{\partial^2}{\partial x^2}\frac{\partial F}{\partial w_{xx}} + \frac{\partial^2}{\partial y^2}\frac{\partial F}{\partial w_{yy}} + \frac{\partial^2}{\partial x \partial y}\frac{\partial F}{\partial w_{xy}} = 0 \tag{4.24}$$

where[5]

$$\frac{\partial F}{\partial w} = -q; \ \frac{\partial F}{\partial w_t} = -I_0\frac{\partial w}{\partial t}; \ \frac{\partial F}{\partial w_{ty}} = -I_1\frac{\partial^2 w}{\partial t \partial y}; \ \frac{\partial F}{\partial w_{tx}} = -I_1\frac{\partial^2 w}{\partial t \partial x};$$

$$\frac{\partial F}{\partial w_{xx}} = (-D^l + Dv)\frac{\partial^2 w}{\partial y^2} + (D + D^l)\frac{\partial^2 w}{\partial x^2};$$

$$\frac{\partial F}{\partial w_{yy}} = (D + D^l)\frac{\partial^2 w}{\partial y^2} - (D^l - Dv)\frac{\partial^2 w}{\partial x^2}$$

$$\frac{\partial F}{\partial w_{xy}} = 2(D + 2D^l - Dv)\frac{\partial^2 w}{\partial x \partial y}; \tag{4.25}$$

Substituting Eqs. (4.25) into Eqs. (4.24) leads to the governing equation:

$$-q + I_0\frac{\partial^2 w}{\partial t^2} - I_1\frac{\partial^4 w}{\partial t^2 \partial y^2} - I_1\frac{\partial^4 w}{\partial t^2 \partial x^2}$$

$$+ (D + D^l)\left(\frac{\partial^4 w}{\partial y^4} + 2\frac{\partial^4 w}{\partial x^2 \partial y^2} + \frac{\partial^4 w}{\partial x^4}\right) = 0 \tag{4.26}$$

which holds on $x \in \leq (0, a)$, $y \in \leq (0, b)$ and $-\frac{h}{2} \leq z \leq \frac{h}{2}$. Meanwhile, the associated flexural rigidity terms are defined as:

$$D = \frac{Eh^3}{12(1 - v^2)}; D^l = \frac{El^2 h}{2(1 + v)} \tag{4.27}$$

Static analysis

For a static analysis, the dependence of the equations on time is suppressed, leading to the following non-homogenous biharmonic equation [14]:

$$(D + D^l)\left(\frac{\partial^4 w}{\partial y^4} + 2\frac{\partial^4 w}{\partial x^2 \partial y^2} + \frac{\partial^4 w}{\partial x^4}\right) = q \tag{4.28}$$

Free vibration analysis

For a free vibration analysis, the transverse load q as well as the terms relating to the rotary inertia are eliminated from the equation to obtain:

---

[5]A change has been made to the sign of the Lagrangian ($\Pi_P - \Pi_K$) to obtain the derivatives in Eq. (4.25)

$$I_0\frac{\partial^2 w}{\partial t^2} + (D + D')\left(\frac{\partial^4 w}{\partial y^4} + 2\frac{\partial^4 w}{\partial x^2 \partial y^2} + \frac{\partial^4 w}{\partial x^4}\right) = 0 \qquad (4.29)$$

which is consistent with the presentation of [15].

4.3 Finite Element Formulations

In this section, a 12-degrees of freedom non-conforming plate element is explored
for the static and vibration analyses of size-dependent microscale plates. As shown
in Fig. 4.1, our consideration is limited to microplates with a rectangular geometrical
domain discretized into series of sub-domain rectangular elements (such as Fig. 4.1b).

For the 12 degrees of freedom plate element, the assumed polynomial functions
for the displacement (w) and the slopes (θ_x and θ_y) in the local coordinate $\xi\eta$ are
[30]:

$$\begin{aligned}
w = {} & A_1 + \xi A_2 + \eta A_3 + \xi^2 A_4 + \eta\xi A_5 + \eta^2 A_6 + \xi^3 A_7 \\
& + \eta\xi^2 A_8 + \eta^2\xi A_9 + \eta^3 A_{10} + \eta\xi^3 A_{11} + \eta^3\xi A_{12}
\end{aligned} \qquad (4.30)$$

$$\begin{aligned}
\theta_x = {} & \frac{dw}{dy} = \frac{1}{b}\frac{dw}{d\eta} \\
= {} & \frac{1}{b}\Big(A_3 + \xi A_5 + 2\eta A_6 + \xi^2 A_8 + 2\eta\xi A_9 \\
& + 3\eta^2 A_{10} + \xi^3 A_{11} + 3\eta^2\xi A_{12}\Big)
\end{aligned} \qquad (4.31)$$

$$\begin{aligned}
\theta_y = {} & \frac{dw}{dx} = -\frac{1}{a}\frac{dw}{d\xi} \\
= {} & -\frac{1}{a}\Big(A_2 + 2\xi A_4 + \eta A_5 + 3\xi^2 A_7 + 2\eta\xi A_8
\end{aligned}$$

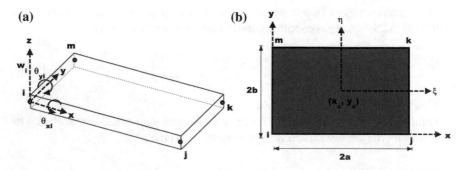

(a)   **(b)**

Fig. 4.1 **a** Schematic of a microplate with nodal degrees of freedom; **b** A rectangular plate element
with both local coordinates $\xi\eta$ and global coordinates xy

$$+\eta^2 A_9 + 3\eta\xi^2 A_{11} + \eta^3 A_{12}\big) \tag{4.32}$$

where $A_1 - A_{12}$ are unknown constants to be found by imposing a set of boundary conditions, specified in Table 4.1, on Eqs. (4.30)–(4.32).

Evaluating Eqs. (4.30)–(4.32) with the boundary conditions yields a system of equations that can be summarized as:

$$
\begin{Bmatrix} w_1 \\ \theta_{x1} \\ \theta_{y1} \\ w_2 \\ \theta_{x2} \\ \theta_{y2} \\ w_3 \\ \theta_{x3} \\ \theta_{y3} \\ w_4 \\ \theta_{x4} \\ \theta_{y4} \end{Bmatrix}
=
\begin{bmatrix}
1 & -1 & -1 & 1 & 1 & 1 & -1 & -1 & -1 & -1 & 1 & 1 \\
0 & 0 & \frac{1}{b} & 0 & -\frac{1}{b} & -\frac{2}{b} & 0 & \frac{1}{b} & \frac{2}{b} & \frac{3}{b} & -\frac{1}{b} & -\frac{3}{b} \\
0 & -\frac{1}{a} & 0 & \frac{2}{a} & \frac{1}{a} & 0 & -\frac{3}{a} & -\frac{2}{a} & -\frac{1}{a} & 0 & \frac{3}{a} & \frac{1}{a} \\
1 & 1 & -1 & 1 & -1 & 1 & 1 & -1 & 1 & -1 & -1 & -1 \\
0 & 0 & \frac{1}{b} & 0 & \frac{1}{b} & -\frac{2}{b} & 0 & \frac{1}{b} & -\frac{2}{b} & \frac{3}{b} & \frac{1}{b} & \frac{3}{b} \\
0 & -\frac{1}{a} & 0 & -\frac{2}{a} & \frac{1}{a} & 0 & -\frac{3}{a} & \frac{2}{a} & -\frac{1}{a} & 0 & \frac{3}{a} & \frac{1}{a} \\
1 & 1 & 1 & 1 & 1 & 1 & 1 & 1 & 1 & 2 & 1 & 1 \\
0 & 0 & \frac{1}{b} & 0 & \frac{1}{b} & \frac{2}{b} & 0 & \frac{1}{b} & \frac{2}{b} & \frac{3}{b} & \frac{1}{b} & \frac{3}{b} \\
0 & -\frac{1}{a} & 0 & -\frac{2}{a} & -\frac{1}{a} & 0 & -\frac{3}{a} & -\frac{2}{a} & -\frac{1}{a} & 0 & -\frac{3}{a} & -\frac{1}{a} \\
1 & -1 & 1 & 1 & -1 & 1 & -1 & 1 & -1 & 1 & -1 & -1 \\
0 & 0 & \frac{1}{b} & 0 & -\frac{1}{b} & \frac{2}{b} & 0 & \frac{1}{b} & -\frac{2}{b} & \frac{3}{b} & -\frac{1}{b} & \frac{3}{b} \\
0 & -\frac{1}{a} & 0 & \frac{2}{a} & -\frac{1}{a} & 0 & -\frac{3}{a} & \frac{2}{a} & -\frac{1}{a} & 0 & -\frac{3}{a} & -\frac{1}{b}
\end{bmatrix}
\begin{Bmatrix} A_1 \\ A_2 \\ A_3 \\ A_4 \\ A_5 \\ A_6 \\ A_7 \\ A_8 \\ A_8 \\ A_8 \\ A_{10} \\ A_{11} \\ A_{12} \end{Bmatrix}
\tag{4.33}
$$

From Eq. (4.33), the vector of unknown coefficients $\{A\}$ are determined as:

$$\{A\} = [H]^{-1}\{w_e\} \tag{4.34}$$

where

Table 4.1 Boundary conditions at the corners of the microplate

Corner 1	Corner 2	Corner 3	Corner 4
$\xi = -1, \eta = -1$: • $w = w_1$ • $\theta_x = \theta_{x1}$ • $\theta_y = \theta_{y1}$	$\xi = 1, \eta = -1$: • $w = w_2$ • $\theta_x = \theta_{x2}$ • $\theta_y = \theta_{y2}$	$\xi = 1, \eta = 1$: • $w = w_3$ • $\theta_x = \theta_{x3}$ • $\theta_y = \theta_{y3}$	$\xi = -1, \eta = 1$: • $w = w_4$ • $\theta_x = \theta_{x4}$ • $\theta_y = \theta_{y4}$

$$
[H]^{-1} = \frac{1}{8}
\begin{bmatrix}
2 & b & -a & 2 & b & a & 2 & -b & a & 2 & -b & -a \\
-3 & -b & a & 3 & b & a & 3 & -b & a & -3 & b & a \\
-3 & -b & a & -3 & -b & -a & 3 & -b & a & 3 & -b & -a \\
0 & 0 & a & 0 & 0 & -a & 0 & 0 & -a & 0 & 0 & a \\
4 & b & -a & -4 & -b & -a & 4 & -b & a & -4 & b & a \\
0 & -b & 0 & 0 & -b & 0 & 0 & b & 0 & 0 & b & 0 \\
1 & 0 & -a & -1 & 0 & -a & -1 & 0 & -a & 1 & 0 & -a \\
0 & 0 & -a & 0 & 0 & a & 0 & 0 & -a & 0 & 0 & a \\
0 & b & 0 & 0 & -b & 0 & 0 & b & 0 & 0 & -b & 0 \\
1 & b & 0 & 1 & b & 0 & -1 & b & 0 & -1 & b & 0 \\
-1 & 0 & a & 1 & 0 & a & -1 & 0 & -a & 1 & 0 & -a \\
-1 & -b & 0 & 1 & b & 0 & -1 & b & 0 & 1 & b & 0
\end{bmatrix} ;
$$

$$
\{w_e\} =
\begin{pmatrix}
w_1 \\
\theta_{x1} \\
\theta_{y1} \\
w_2 \\
\theta_{x2} \\
\theta_{y2} \\
w_3 \\
\theta_{x3} \\
\theta_{y3} \\
w_4 \\
\theta_{x4} \\
\theta_{y4}
\end{pmatrix}
\tag{4.35}
$$

In the light of Eq. (4.35), the assumed transverse displacement and rotation functions become:

$$
w = \begin{bmatrix} 1 & \xi & \eta & \xi^2 & \eta\xi & \eta^2 & \xi^3 & \eta\xi^2 & \eta^2\xi & \eta^3 & \eta\xi^3 & \eta^3\xi \end{bmatrix}
$$
$$
\{A\} = \underline{\begin{bmatrix} 1 & \xi & \eta & \xi^2 & \eta\xi & \eta^2 & \xi^3 & \eta\xi^2 & \eta^2\xi & \eta^3 & \eta\xi^3 & \eta^3\xi \end{bmatrix}[H]^{-1}\{w_e\}}
\tag{4.36}
$$

The interpolation functions are obtained by evaluating the underlined terms in Eq. (4.36), leading to:

$$
w(\xi, \eta, t) = [N_w(\xi, \eta)]\{w_e\}
$$
$$
= \big[N_{w1} \; N_{w2} \; N_{w3} \; N_{w4} \; N_{w5} \; N_{w6} \; N_{w7}
$$
$$
N_{w8} \; N_{w9} \; N_{w10} \; N_{w11} \; N_{w12} \big]\{w_e\}
\tag{4.37}
$$

The full form of the interpolation functions are provided in Eq. (4.38).

$$N_w^T = \begin{bmatrix} -\dfrac{1}{8}(-1+\eta)(-1+\xi)(-2+\eta+\eta^2+\xi+\xi^2) \\[2mm] -\dfrac{1}{8}b(-1+\eta)^2(1+\eta)(-1+\xi) \\[2mm] \dfrac{1}{8}a(-1+\eta)(-1+\xi)^2(1+\xi) \\[2mm] \dfrac{1}{8}(-1+\eta)(1+\xi)(-2+\eta+\eta^2-\xi+\xi^2) \\[2mm] \dfrac{1}{8}b(-1+\eta)^2(1+\eta)(1+\xi) \\[4mm] \dfrac{1}{8}a(-1+\eta)(-1+\xi)(1+\xi)^2 \\[2mm] -\dfrac{1}{8}(1+\eta)(1+\xi)(-2-\eta+\eta^2-\xi+\xi^2) \\[2mm] \dfrac{1}{8}b(-1+\eta)(1+\eta)^2(1+\xi) \\[2mm] -\dfrac{1}{8}a(1+\eta)(-1+\xi)(1+\xi)^2 \\[2mm] \dfrac{1}{s}(1+\eta)(-1+\xi)(-2-\eta+\eta^2+\xi+\xi^2) \\[2mm] -\dfrac{1}{8}b(-1+\eta)(1+\eta)^2(-1+\xi) \\[4mm] -\dfrac{1}{8}a(1+\eta)(-1+\xi)^2(1+\xi) \end{bmatrix} \tag{4.38}$$

4.3.1 Stiffness Matrix

Using Eq. (4.38), the curvature matrix is obtained as:

$$\{\kappa\} = \left\{ \begin{array}{c} \kappa_x \\ \kappa_y \\ \kappa_{xy} \end{array} \right\} = \left\{ \begin{array}{c} -\dfrac{\partial^2 w}{\partial x^2} \\ -\dfrac{\partial^2 w}{\partial y^2} \\ \dfrac{\partial^2 w}{\partial x \partial y} \end{array} \right\} = [B]\{w_e\} \tag{4.39}$$

where $[B]$ is a 3×12 matrix and its transpose is given by:

$$[\boldsymbol{B}]^T = \begin{bmatrix}
\dfrac{3(-1+n)\xi}{4a^2} & \dfrac{3n(-1+\xi)}{4b^2} & \dfrac{-4+3\eta^2+3\xi^2}{4ab} \\[2mm]
0 & \dfrac{(-1+3n)(-1+\xi)}{4b} & -\dfrac{1+2\eta-3n^2}{4a} \\[2mm]
\dfrac{-1+n+3\xi-3n\xi}{4a} & 0 & \dfrac{1+2\xi-3\xi^2}{4b} \\[2mm]
-\dfrac{3(-1+\eta)\xi}{4a^2} & -\dfrac{3\eta(1+\xi)}{4b^2} & \dfrac{4-3n^2-3\xi^2}{4ab} \\[2mm]
0 & -\dfrac{(-1+3\eta)(1+\xi)}{4b} & \dfrac{1+2n-3n^2}{4a} \\[2mm]
-\dfrac{(-1+\eta)(1+3\xi)}{4a} & 0 & \dfrac{1-2\xi-3\xi^2}{4b} \\[2mm]
\dfrac{3(1+\eta)\xi}{4a^2} & \dfrac{3\eta(1+\xi)}{4b^2} & \dfrac{-4+3n^2+3\xi^2}{4ab} \\[2mm]
0 & -\dfrac{(1+3\eta)(1+\xi)}{4b} & \dfrac{1-2n-3n^2}{4a} \\[2mm]
\dfrac{(1+\eta)(1+3\xi)}{4a} & 0 & \dfrac{-1+2\xi+3\xi^2}{4b} \\[2mm]
-\dfrac{3(1+\eta)\xi}{4a^2} & -\dfrac{3n(-1+\xi)}{4b^2} & \dfrac{4-3n^2-3\xi^2}{4ab} \\[2mm]
0 & \dfrac{(1+3n)(-1+\xi)}{4b} & \dfrac{-1+2n+3n^2}{4a} \\[2mm]
\dfrac{(1+\eta)(-1+3\xi)}{4a} & 0 & -\dfrac{1+2\xi-3\xi^2}{4b}
\end{bmatrix} \tag{4.40}$$

In a similar spirit, using Eq. (4.38), the symmetric curvature matrix is obtained as:

$$\left\{ \begin{array}{c} s_x \\ s_y \\ s_{xy} \end{array} \right\} = \left\{ \begin{array}{c} \dfrac{\partial^2 w}{\partial x \partial y} \\[2mm] -\dfrac{\partial^2 w}{\partial x \partial y} \\[2mm] \dfrac{\partial^2 w}{\partial y^2} - \dfrac{\partial^2 w}{\partial x^2} \end{array} \right\} = [\boldsymbol{B}_c]\{w_e\} \tag{4.41}$$

where $[B_c]$ is a 3×12 matrix whose transpose is given by:

$$[\boldsymbol{B}_c]^T = \begin{bmatrix}
\dfrac{4-3n^2-3\xi^2}{8ab} & \dfrac{-4+3\eta^2+3\xi^2}{8ab} & \dfrac{3\big(b^2(-1+\eta)\xi+a^2(n-n\xi)\big)}{4a^2b^2} \\[2mm]
\dfrac{1+2\eta-3\eta^2}{8a} & -\dfrac{1+2\eta-3\eta^2}{8a} & \dfrac{-1-3n(-1+\xi)+\xi}{4b} \\[2mm]
-\dfrac{1+2\xi-3\xi^2}{8b} & \dfrac{1+2\xi-3\xi^2}{8b} & \dfrac{-1+n+3\xi-3n\xi}{4a} \\[2mm]
\dfrac{-4+3\eta^2+3\xi^2}{8ab} & \dfrac{4-3n^2-3\xi^2}{8ab} & \dfrac{3\big(-b^2(-1+\eta)\xi+a^2n(1+\xi)\big)}{4a^2b^2} \\[2mm]
-\dfrac{1+2n-3n^2}{8a} & \dfrac{1+2\eta-3\eta^2}{8a} & \dfrac{(-1+3\eta)(1+\xi)}{4a} \\[2mm]
\dfrac{-1+2\xi+3\xi^2}{8b} & \dfrac{1-2\xi-3\xi^2}{8b} & -\dfrac{(-1+\eta)(1+3\xi)}{4a} \\[2mm]
\dfrac{4-3n^2-3\xi^2}{8ab} & \dfrac{-4+3n^2+3\xi^2}{8ab} & \dfrac{3b^2(1+\eta)\xi-3a^2\eta(1+\xi)}{4a^2b^2} \\[2mm]
\dfrac{-1+2n+3n^2}{8a} & \dfrac{1-2\eta-3\eta^2}{8a} & \dfrac{(1+3\eta)(1+\xi)}{4b} \\[2mm]
\dfrac{1-2\xi-3\xi^2}{8b} & \dfrac{-1+2\xi+3\xi^2}{8b} & \dfrac{(1+\eta)(1+3\xi)}{4a} \\[2mm]
\dfrac{-4+3n^2+3\xi^2}{8ab} & \dfrac{4-3n^2-3\xi^2}{8ab} & -\dfrac{3\big(b^2(1+\eta)\xi+a^2(n-n\xi)\big)}{4a^2b^2} \\[2mm]
\dfrac{1-2n-3n^2}{8a} & \dfrac{-1+2n+3n^2}{8a} & \dfrac{(1+3n)(-1+\xi)}{4b} \\[2mm]
\dfrac{1+2\xi-3\xi^2}{8b} & -\dfrac{1+2\xi-3\xi^2}{8b} & \dfrac{(1+\eta)(-1+3\xi)}{4a}
\end{bmatrix} \tag{4.42}$$

It is now possible to obtain the stiffness matrix as:

$$\frac{\partial \Pi_s}{\partial \{w_e\}^T} = 0 = \left(\left(\frac{Eh^3}{12(1-v^2)} \right) ab \left\{ \int\limits_{-1}^{1} \int\limits_{-1}^{1} [B]^T [C_1][B] d\eta d\xi \right\} \right)$$

$$+ \left(\frac{El^2h}{2(1+v)} \right) ab \left\{ \int_{-1}^{1} \int_{-1}^{1} [\boldsymbol{B}_c]^T [C_2][\boldsymbol{B}_c] d\eta d\xi \right\} \right) - \{f_e\} \quad (4.43)$$

$$[C_1] = \begin{bmatrix} 1 & v & 0 \\ v & 1 & 0 \\ 0 & 0 & \frac{1-v}{2} \end{bmatrix}; [C_2] = \begin{bmatrix} 1 & 0 & 0 \\ 0 & 1 & 0 \\ 0 & 0 & 1 \end{bmatrix} \quad (4.44)$$

Consequently, the matrix equation of the microscale plate element is obtained as:

$$\{f_e\} = [K_P]\{w_e\} \quad (4.45)$$

where

$$\{f_e\} = \begin{Bmatrix} f_{w1} \\ m_{x1} \\ m_{y1} \\ f_{w2} \\ m_{x2} \\ m_{y2} \\ f_{w3} \\ m_{x3} \\ m_{y3} \\ f_{w4} \\ m_{x4} \\ m_{y4} \end{Bmatrix} \quad (4.46)$$

$$[K_P] = [K_c] + [K_{mcst}] \quad (4.47)$$

Motivated by the idea in Petyt [31], the entries of the 12 by 12 stiffness-related matrices, $[K_c]$ and $[K_{mcst}]$ in Eq. (4.47), are broken down into series of 3 by 3 submatrices. That is:

$$[K_c] = \frac{D}{30b^2r^3} \begin{bmatrix} k_{11} & k_{12} & k_{13} & k_{14} \\ k_{21} & k_{22} & k_{23} & k_{24} \\ k_{31} & k_{32} & k_{33} & k_{34} \\ k_{41} & k_{42} & k_{43} & k_{44} \end{bmatrix} \quad (4.48)$$

$$[K_{mcst}] = \frac{Ehs^2}{30b^2r^3(1+v)} \begin{bmatrix} s_{11} & s_{12} & s_{13} & s_{14} \\ s_{21} & s_{22} & s_{23} & s_{24} \\ s_{31} & s_{32} & s_{33} & s_{34} \\ s_{41} & s_{42} & s_{43} & s_{44} \end{bmatrix} \quad (4.49)$$

$$k_{11} = \begin{bmatrix} 30 + 30r^4 + r^2(21 - 6v) & 3br^2(1 + 10r^2 + 4v) & -3br(10 + r^2(1 + 4v)) \\ 3br^2(1 + 10r^2 + 4v) & 8b^2r^2(1 + 5r^2 - v) & -30b^2r^3v \\ -3br(10 + r^2(1 + 4v)) & -30b^2r^3v & 8b^2r^2(5 - r^2(-1 + v)) \end{bmatrix} \quad (4.50)$$

$$k_{21} = k_{12} \begin{bmatrix} 3\left(-10 + 5r^4 + r^2(-7 + 2v)\right) & 3br^2\left(-1 + 5r^2 - 4v\right) & 3br\left(10 - r^2(-1 + v)\right) \\ 3br^2\left(-1 + 5r^2 - 4v\right) & 4b^2r^2\left(-2 + 5r^2 + 2v\right) & 0 \\ 3br\left(-10 + r^2(-1 + v)\right) & 0 & 2b^2r^2\left(10 + r^2(-1 + v)\right) \end{bmatrix}$$

(4.51)

$$k_{31} = k_{13} \begin{bmatrix} -3\left(5 + 5r^4 + r^2(-7 + 2v)\right) & -3br^2\left(-1 + 5r^2 + v\right) & 3br\left(5 + r^2(-1 + v)\right) \\ 3br^2\left(-1 + 5r^2 + v\right) & 2b^2r^2\left(1 + 5r^2 - v\right) & 0 \\ 3br\left(-5 - r^2(-1 + v)\right) & 0 & 2b^2r^2\left(5 - r^2(-1 + v)\right) \end{bmatrix}$$

(4.52)

$$k_{41} = k_{14} \begin{bmatrix} 15 - 30r^4 + 3r^2(-7 + 2v) & 3br^2\left(-1 - 10r^2 + v\right) & 3br\left(-5 + r^2(1 + 4v)\right) \\ 3br^2\left(1 + 10r^2 - v\right) & 2b^2r^2\left(-1 + 10r^2 + v\right) & 0 \\ 3br\left(-5 + r^2(1 + 4v)\right) & 0 & 4b^2r^2\left(5 + 2r^2(-1 + v)\right) \end{bmatrix}$$

(4.53)

$$s_{11} = \begin{bmatrix} 15\left(2 - r^2 + 2r^4\right) & 15br^2\left(-1 + 2r^2\right) & 15br\left(-2 + r^2\right) \\ 15br^2\left(-1 + 2r^2\right) & 40b^2r^4 & 30b^2r^3 \\ 15br\left(-2 + r^2\right) & 30b^2r^3 & 40b^2r^2 \end{bmatrix}$$

(4.54)

$$s_{21} = \begin{bmatrix} 15\left(-2 + r^2 + r^4\right) & 15br^2\left(1 + r^2\right) & 30br \\ 15br^2\left(1 + r^2\right) & 20b^2r^4 & 0 \\ -30br & 0 & 20b^2r^2 \end{bmatrix}$$

(4.55)

$$s_{31} = \begin{bmatrix} -15\left(1 + r^2 + r^4\right) & -15br^4 & 15br \\ 15br^4 & 10b^2r^4 & 0 \\ -15br & 0 & 10b^2r^2 \end{bmatrix}$$

(4.56)

$$s_{41} = \begin{bmatrix} 15\left(1 + r^2 - 2r^4\right) & -30br^4 & -15br\left(1 + r^2\right) \\ 30br^4 & 20b^2r^4 & 0 \\ -15br\left(1 + r^2\right) & 0 & 20b^2r^2 \end{bmatrix}$$

(4.57)

where $r = b/a$. The other contents of the classical stress-related stiffness matrix and the couple stress-related stiffness matrix are listed in Table 4.2.

Table 4.2 Elements of the stiffness matrix

	Elements of K_c	Elements of K_{mcst}
1	$k_{22} = I_3^T k_{11} I_3$	$s_{22} = I_3^T s_{11} I_3$
2	$k_{32} = I_3^T k_{41} I_3$	$s_{32} = I_3^T s_{41} I_3$
3	$k_{33} = I_1^T k_{11} I_1$	$s_{33} = I_1^T s_{11} I_1$
4	$k_{42} = I_3^T k_{31} I_3$	$s_{42} = I_3^T s_{31} I_3$
5	$k_{43} = I_1^T k_{21} I_1$	$s_{43} = I_1^T s_{21} I_1$
6	$k_{44} = I_2^T k_{21} I_2$	$s_{44} = I_2^T s_{21} I_2$

4.3.2 Equivalent Load Vector

For a rectangular plate subjected to a uniformly distributed surface pressure $q\left(\text{N/m}^2\right)$, the vector of equivalent load is obtained from:

$$q_{ev} = ab\left\{ \int_{-1}^{1}\int_{-1}^{1}[N_w]^T q[N_w]d\eta d\xi \right\} \tag{4.58}$$

When evaluated, Eq. (4.58) produces a column matrix:

$$\{q_{ev}\} = q\frac{ab}{3}\left[3\ b\ -a\ 3\ b\ a\ 3\ -b\ a\ 3\ -b\ -a \right]^T \tag{4.59}$$

4.3.3 Mass Matrix

By looking at the strong form of the governing equation of motion in Eq. (4.29), it is apparent that the mass matrix can be obtained as:

$$[M_P] = \rho hab\left\{ \int_{-1}^{1}\int_{-1}^{1}[N_w]^T[N_w]d\eta d\xi \right\} \tag{4.60}$$

$$[M_P] = \frac{\rho hab}{3150}\begin{bmatrix} m_{11} & m_{12} \\ m_{12}^T & m_{22} \end{bmatrix} \tag{4.61}$$

$$m_{11} = \begin{bmatrix} 1727 & 461b & -461a & 613 & 199b & 274a \\ 461b & 160b^2 & -126ab & 199b & 80b^2 & 84ab \\ -461a & -126ab & 160a^2 & -274a & -84ab & -120a^2 \\ 613 & 199b & -274a & 1727 & 461b & 461a \\ 199b & 80b^2 & -84ab & 461b & 160b^2 & 126ab \\ 274a & 84ab & -120a^2 & 461a & 126ab & 160a^2 \end{bmatrix} \tag{4.62}$$

$$m_{12} = \begin{bmatrix} 197 & -116b & 116a & 613 & -274b & -199a \\ 116b & -60b^2 & 56ab & 274b & -120b^2 & -84ab \\ -116a & 56ab & -60a^2 & -199a & 84ab & 80a^2 \\ 613 & -274b & 199a & 197 & -116b & -116a \\ 274b & -120b^2 & 84ab & 116b & -60b^2 & -56ab \\ 199a & -84ab & 80a^2 & 116a & -56ab & -60a^2 \end{bmatrix} \tag{4.63}$$

$$m_{22} = \begin{bmatrix} 1727 & -461b & 461a & 613 & -199b & -274a \\ -461b & 160b^2 & -126ab & -199b & 80b^2 & 84ab \\ 461a & -126ab & 160a^2 & 274a & -84ab & -120a^2 \\ 613 & -199b & 274a & 1727 & -461b & -461a \\ -199b & 80b^2 & -84ab & -461b & 160b^2 & 126ab \\ -274a & 84ab & -120a^2 & -461a & 126ab & 160a^2 \end{bmatrix} \tag{4.64}$$

Consequently, the discretized time-dependent equations governing the free oscillatory motion of the microstructure-dependent plate takes the form:

$$M_P\left(\frac{d^2 w_e(t)}{dt^2}\right) + K_P w_e(t) = 0 \tag{4.65}$$

Following the same steps as in Chap. 3, we set up an eigenvalue problem of the form

$$\left[-\omega^2 M_P + K_P\right] w_e e^{i\omega t} = 0 \tag{4.66}$$

The desired natural frequencies of the plate element are then sought via a nontrivial solution search in the form:

$$\left| K_P - \omega^2 M_P \right| = 0 \tag{4.67}$$

4.4   R Functions for Static and Vibration Analyses

The primary functions developed for this chapter can be used to evaluate the size-dependent stiffness matrix, the mass matrix and the vector of equivalent uniformly distributed load of the microplate element. Along with the aforementioned are secondary functions that can be used to apply boundary conditions on the global matrix of a system of microplate elements. Again, all the functions are included in **microfiniteR**. To avoid an unnecessary crowding of the pages with hundreds of line of codes, only the names of the R functions developed for this chapter are listed in Table 4.3. Readers are encouraged to check the book website for the complete listing of the codes.

4.5   Bending and Free Vibration Analyses with the Implemented R Functions

Example 4.1 Bending of a fully clamped square plate with no microstructure-dependent property

Problem Compute the central deflection of a clamped square plate subjected to a uniformly distributed load using symmetry boundary conditions. The following properties are used for computational purpose[6]: $E = 10920\,psi$; $v = 0.3$; $L = 1''$; $h = 0.1''$; $q = 1$psi.

Table 4.3 Highlight of implemented R functions for micro Kirchhoff plate element (MKP)

Functions for evaluating element's matrix and vectors

Functions	Parameters
FormStiffnessMKplate(youngmod, poissonratio, edge_a, edge_b, thickness,lengthscale)	• Young's modulus • Poisson's ratio • Length along x • Length along y • Thickness • Material length-scale
FormMassMKplate(youngmod, edge_a, edge_b, thickness, rho)	• Young's modulus • Length along x • Length along y • Mass density
ExpandStiffnessMKPlate (tdof, eMatrix, i, j, k, m)	• Total degrees of freedom • Element's stiffness matrix • Nodal indices (i, j, k, m)
`FormLoadMKPlate` (pressuremag, edge_a, edge_b)	• Magnitude of uniform pressure load • Length along x • Length along y
ExpandLoadMKPlate(tdof, elementloadMatrix, i, j, k, m)	• Total degrees of freedom • Element's Load matrix • Nodal indices (i, j, k, m)
`FormReducedLoad2D` (bigColumnMatrix,unrestrainednodes)	• Global vector of equivalent loads • Unrestrained nodes

Functions for boundary conditions

`FixNodes2D`(nodes) `HingeNodes2Dx`(nodes) `HingeNodes2Dy`(nodes) `Symmetry2Dy`(nodes) `Symmetry2Dx`(nodes)	Index of a single node or a set of indices of nodes that are restrained.

Solution

A common practice in the analysis of plate problems is to employ symmetry, if it exists (and it does for many simple loading and boundary condition such as happen in the current case) [30, 32]. However, the current solution does not employ symmetry[7] (see next example). Instead, the plate is discretized into four subdomains as shown in Fig. 4.2b.

Further, as indicated in Sect. 4.3, the developed plate element is characterized by dimensions $2a \times 2b$. Therefore, if the side of each element is 0.5″, then $2a = 0.5″$ or $a = b = 0.25″$. The complete listing of codes for the solution is shown next.

---

[6]The geometric and loading properties specified here are not necessarily that of a microplate, but have been used to allow for validation with published results.

[7]The use of symmetry is much more efficient than solving the entire domain.

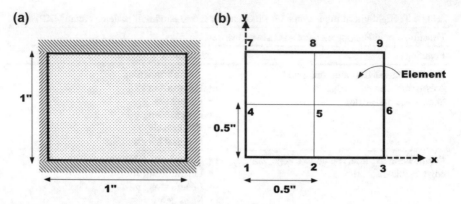

Fig. 4.2 a A clamped square plate; **b** discretized subdomains

Step 1   Load the **microfiniteR** package, supply the material/geometric properties of the plate and declare the number of elements.

```
library(microfiniteR)
eE = 10920                                    # Young's modulus
edgey = 1/4                                    #Element edge length along y
edge_ratio = 1
edgex = edge_ratio * edgey
pratio = 0.3
height = 0.1                                    #Thickness
q = 1
lscale = 0.0 * height                           #Material length-scale parameter = 0

enum = 4;                                       #Number of element
numNodes = 9;
dofpernode = 3;                                 #Degrees of freedom per node
tdof = (numNodes)*dofpernode;

# Elements nodal indices
e1 = c(1, 2, 5, 4)                              #Index of element 1
e2 = c(2, 3, 6, 5)
e3 = c(5, 6, 9, 8)
e4 = c(4, 5, 8, 7)
```

Step 2   Form   the   elements   stiffness   matrices   using   the   function
FormStiffnessMKplate() and then establish the global stiffness
matrix.

```
k1 = FormStiffnessMKplate(eE, pratio, edgex, edgey, height, lscale); k1

k2 = k1;

k3 = k1;

k4 = k1;


# Expand stiffness matrices towards forming the global  matrix

K1 = ExpandStiffnessMKPlate(tdof, k1, 1, 2, 5, 4); K1

K2 = ExpandStiffnessMKPlate(tdof, k2, 2, 3, 6, 5);

K3 = ExpandStiffnessMKPlate(tdof, k3, 5, 6, 9, 8);

K4 = ExpandStiffnessMKPlate(tdof, k4, 4, 5, 8, 7);

bigK = K1 + K2 + K3 + K4
```

Each of the first four lines computes the stiffness matrix for each element, and these are then fed into **ExpandStiffnessMKPlate()**. The global stiffness matrix is stored in the variable **bigK**.

Step 3   Form the vector of equivalent load for each element as well as for the global system.

```
# Elements load vectors

e1_load = FormLoadMKPlate(q, edgex, edgey); e1_load

e2_load = e1_load

e3_load = e1_load

e4_load = e1_load


# Expand load vector

bigL1 = ExpandLoadMKPlate(tdof, e1_load, 1, 2, 5, 4);

bigL2 = ExpandLoadMKPlate(tdof, e2_load, 2, 3, 6, 5);

bigL3 = ExpandLoadMKPlate(tdof, e3_load, 5, 6, 9, 8);

bigL4 = ExpandLoadMKPlate(tdof, e4_load, 4, 5, 8, 7);

bigL = bigL1 + bigL2 + bigL3 + bigL4
```

Step 4   Apply the boundary condition(s) on the global stiffness to obtain the corresponding reduced matrices.

For the boundary conditions, all the nodes except node 5 are fixed. We call the function **FixNodes2D()** on these nodes to eliminate the vertical and rotational displacements. Further, we used **ExtractFreeRows()** to obtain the pointers to the degrees of freedom of the unrestrained nodes. Finally, the output of **ExtractFreeRows()** is used with **FormReducedMatrix()** and **FormReducedLoad2D()** to establish the reduced stiffness matrix and the reduced load vector, respectively.

```
restrained_nodes = FixNodes2D(c(1:4, 6:9));

freenodes = ExtractFreeRows(tdof, restrained_nodes);freenodes


reducedK = FormReducedMatrix(bigK, freenodes);

reducedL = FormReducedLoad2D(bigL, freenodes); reducedLoad
```

The reduced stiffness matrix and the reduced load vector are obtained as:

```
> reducedK = FormReducedMatrix(bigK, freenodes);reducedK
        [,1]  [,2]  [,3]
[1,]  168.96 0.00  0.00
[2,]    0.00 6.08  0.00
[3,]    0.00 0.00  6.08
> reducedLoad = FormReducedLoad2D(bigL, freenodes); reducedLoad
      reducedloadvector
[1,]              0.25
[2,]              0.00
[3,]              0.00
```

Step 5   Find the unknown displacements

```
unknowndisp = FindNodalDOFs(reducedK,reducedLoad); unknowndisp
```

The above produces the following output:

```
> unknowndisp = FindNodalDOFs(reducedK,reducedLoad); unknowndisp
      reducedloadvector
[1,]         0.00147964
[2,]         0.00000000
[3,]         0.00000000
```

We obtained a value of 0.00147 for the deflection, which is about 14% more than 0.0013 predicted by the *Method of Fundamental Solution* employed by Tsiatas [14] (see column 5 of Table 1 in [14]). Meanwhile, the exact solution for the central deflection of a clamped square plate with no microstructure-dependent property is given by $w_c = 0.00126qa^4/D$ [33]. A difference of 17% is noted with the exact solution. In the next example, we will employ symmetry to get even close to the result of Tsiatas [14].

Example 4.2 Bending of a clamped square plate with no microstructure-dependent property considering symmetry

Problem Compute the central deflection of a clamped square plate subjected to a uniformly distributed load.

The following properties are used for computational purpose: $E = 10920psi; v = 0.3; L = 1''; h = 0.1''; q = 1psi$.

Solution

Since the plate is doubly symmetric, one-quarter of the plate can be employed for faster computation of the results as shown in Fig. 4.3. The quarter plate has edge dimensions $0.5'' \times 0.5''$ as indicated in Fig. 4.3b. For improved results, the quarter plate is discretized into a 4 by 4 subdomain containing a total of 16 elements. Consequently, the side of each element is $(1/8)''$ and $a = b = (1/16)''$. We employed this value and the data provided in the problem statement in the code snippets that follow.

Step 1 Load the **microfiniteR** package, supply the material/geometric properties of the plate and declare the number of elements.

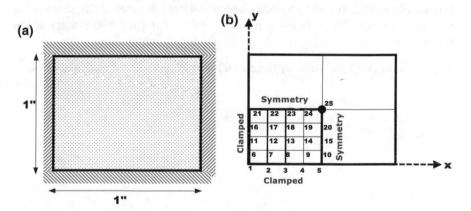

Fig. 4.3 **a** A clamped square plate; **b** a 4×4 discretized subdomains

```
library(microfiniteR)

eE = 10920

edgey = 1/16                        #The new element edge dimension

edge_ratio = 1

edgex = edge_ratio * edgey

pratio = 0.3

height = 0.1

q = 1


lenscale = 0*height                 #Eliminate the effect of material length-scale

enum = 16;                          #Number of element

numNodes = 25;                      #Number of nodes

dofpernode = 3;

tdof = (numNodes)*dofpernode;
```

In the next step (and the rest of the chapter), we will skip the explicit process of establishing the elements characteristics matrices and vectors (equivalent load). Instead, these quantities will be computed implicitly by using the appropriate functions that have been developed to establish the global properties. For instance **FormGlobalKM2D()** and **FormGlobalLoad2D()** are used in this sense in the next line of codes. These function require the data about the connectivity of the nodes.

Step 2   Form the global stiffness matrix and the global vector of equivalent load.

```
# Elements nodal indices

e1 = c(1, 2, 7, 6)                    #Nodal indices of element 1

e2 = c(2, 3, 8, 7)

e3 = c(3, 4, 9, 8)

e4 = c(4, 5, 10, 9)

e5 = c(6, 7, 12, 11)

e6 = c(7, 8, 13, 12)

e7 = c(8, 9, 14, 13)

e8 = c(9, 10, 15, 14)


e9 = c(11, 12, 17, 16)

e10 = c(12, 13, 18, 17)

e11 = c(13, 14, 19, 18)

e12 = c(14, 15, 20, 19)

e13 = c(16, 17, 22, 21)

e14 = c(17, 18, 23, 22)

e15 = c(18, 19, 24, 23)

e16 = c(19, 20, 25, 24)          #Nodal indices of element 1


# We put the nodes in a list to iterate over each one of them
list_nodes=list(e1, e2, e3, e4,
                e5, e6, e7, e8,
                e9, e10, e11, e12,
                e13, e14, e15, e16)
#Global stiffness matrix
bigK <- FormGlobalKM2D(enum, list_nodes, eE, pratio,
                       edgex, edgey, height, lenscale, 0, 1)


# Global load vector
bigL <- FormGlobalLoad2D(enum, list_nodes, q, edgex, edgey)
```

Step 3 Apply the boundary condition(s) on the global matrices.

For the boundary conditions, nodes 1 to 6 as well as 11, 16 & 21 are fixed. Therefore, we will call the function **FixNodes2D()** on these nodes to eliminate the vertical and rotational displacements. Symmetry boundary conditions are applied to the nodes along the lines of symmetry (at the right-side and top edges of the quarter plate). The functions **Symmetry2Dy()** and **Symmetry2Dx()** are developed for the purpose of applying the symmetry boundary conditions.

```
# Specify the restrained nodes
restrained_nodesABAD = FixNodes2D(c(1:5, 6, 11, 16, 21));
restrained_nodesBC = Symmetry2Dy(c(10, 15, 20, 25));
restrained_nodesCD = Symmetry2Dx(c(22:25));


# Combine the restrained nodes
restrained_nodes = unique(c(restrained_nodesABAD, restrained_nodesBC,
                            restrained_nodesCD));


# Obtain the free nodes
freenodes = ExtractFreeRows(tdof, restrained_nodes);freenodes
```

The output of the last line is printed to reveal two things: (i) the total number of degrees of freedom that will participate in the computation; and (ii) the position of the deflection of node 25 (which is located at the center of the plate). Notice that the 40th element in the vector stored in variable **freenodes** correspond to the displacement of node 25.

```
> freenodes = ExtractFreeRows(tdof, restrained_nodes);freenodes
 [1] 19 20 21 22 23 24 25 26 27 28 29 34 35 36 37 38 39 40 41 42 43 44 49 50 51 52 53 54
55 56 57 58 59 64 66 67
[37] 69 70 72 73
```

The reduced stiffness matrix (40 by 40) and the reduced load vector (40 by 1) are obtained as:

```
reducedK = FormReducedMatrix(bigK, freenodes);
reducedLoad = FormReducedLoad2D(bigL, freenodes);
```

Step 4   Find the unknown displacements

```
unknowndisp = FindNodalDOFs(reducedK,reducedLoad);
unknowndisp[40]
```

The above produces the following output:

```
> unknowndisp = FindNodalDOFs(reducedK,reducedLoad); unknowndisp[40]
[1] 0.001303946
```

Here the 16-element solution produced a value of 0.001303 for the deflection, which matches the prediction of the *Method of Fundamental Solution* employed by Tsiatas [14]. In the next example, we will look at the effect of the material length-scale parameter on the bending response.

Example 4.3 Bending of a clamped square plate with microstructure-dependent property considering symmetry

Problem Examine the effect of the material length-scale on the bending response of a clamped square plate subjected to a uniformly distributed load.

The following properties are used for computational purpose: $E = 10920\ psi$; $v = 0.3$; $L = 1''$; $h = 0.1''$; $q = 1\ psi$.

Solution
We will approach the question in a manner similar to the previous example. That is, employ symmetry and discretize the solution domain into 16 elements. The code snippet is provided below. The presence of the material length-scale will be noticed in line 9 of the code below.

```
library(microfiniteR)

eE = 10920

edgey = 1/16                       #The new element edge dimension

edge_ratio = 1

edgex = edge_ratio * edgey

pratio = 0.3

height = 0.1

q = 1


lenscale = 0.3*height              #Include the effect of material lengthscale

enum = 16;                         #Number of element

numNodes = 25;

dofpernode = 3;

tdof = (numNodes)*dofpernode;


# Elements nodal indices
e1 = c(1, 2, 7, 6)                 #Nodal indices of element 1

e2 = c(2, 3, 8, 7)
```

```
e3 = c(3, 4, 9, 8)
e4 = c(4, 5, 10, 9)
e5 = c(6, 7, 12, 11)
e6 = c(7, 8, 13, 12)
e7 = c(8, 9, 14, 13)
e8 = c(9, 10, 15, 14)


e9 = c(11, 12, 17, 16)
e10 = c(12, 13, 18, 17)
e11 = c(13, 14, 19, 18)
e12 = c(14, 15, 20, 19)
e13 = c(16, 17, 22, 21)
e14 = c(17, 18, 23, 22)
e15 = c(18, 19, 24, 23)
e16 = c(19, 20, 25, 24)          #Nodal indices of element 1

# We put the nodes in a list to iterate over each one of them
list_nodes=list(e1, e2, e3, e4,
                e5, e6, e7, e8,
                e9, e10, e11, e12,
                e13, e14, e15, e16)

#Global stiffness matrix
bigK <- FormGlobalKM2D(enum, list_nodes, eE, pratio,
                       edgex, edgey, height, lenscale, 0, 1)

# Global load vector
bigL <- FormGlobalLoad2D(enum, list_nodes, q, edgex, edgey)

# Specify the restrained nodes
restrained_nodesABAD = FixNodes2D(c(1:5, 6, 11, 16, 21));
restrained_nodesBC = Symmetry2Dy(c(10, 15, 20, 25));
restrained_nodesCD = Symmetry2Dx(c(22:25));

# Combine the restrained nodes
restrained_nodes = unique(c(restrained_nodesABAD, restrained_nodesBC,
                            restrained_nodesCD));



# Obtain the free nodes
freenodes = ExtractFreeRows(tdof, restrained_nodes);freenodes


reducedK = FormReducedMatrix(bigK, freenodes);
reducedLoad = FormReducedLoad2D(bigL, freenodes);


unknowndisp = FindNodalDOFs(reducedK,reducedLoad);
unknowndisp[40]
```

By running the preceding code snippet, the output below is obtained:

```
> unknowndisp[40]
[1] 0.0009277956
```

For a material length-scale parameter that is 1/3 of the thickness, the computation produces a deflection value of 0.00092. This again matches what is predicted by Tsiatas [14] (see column 4 of Table 3 in [14]).

Example 4.4 Bending of a partially simply-supported square plate with microstructure-dependent property considering symmetry

Problem Examine the effect of the material length-scale parameter on the bending response of a square plate subjected to a uniformly distributed load with simply-supported boundary conditions at $x = 0$ and $x = 1''$.

The following properties are used for computational purpose: $E = 10920\,psi; v = 0.3; L = 1''; h = 0.1''; q = 1\,psi; h/l = 0.3$.

Solution
The code snippet for this problem is provided below.

```
library(microfiniteR)
eE = 10920
edgey = 1/16                        #The new element edge dimension
edge_ratio = 1
edgex = edge_ratio * edgey
pratio = 0.3
height = 0.1
q = 1


lenscale = (0.3)*height             #Include the effect of material length-scale
enum = 16;                          #Number of element
numNodes = 25;
dofpernode = 3;
tdof = (numNodes)*dofpernode;

# Elements nodal indices
```

```
e1 = c(1, 2, 7, 6)                       #Nodal indices of element 1
e2 = c(2, 3, 8, 7)
e3 = c(3, 4, 9, 8)
e4 = c(4, 5, 10, 9)
e5 = c(6, 7, 12, 11)
e6 = c(7, 8, 13, 12)
e7 = c(8, 9, 14, 13)
e8 = c(9, 10, 15, 14)


e9 = c(11, 12, 17, 16)
e10 = c(12, 13, 18, 17)
e11 = c(13, 14, 19, 18)
e12 = c(14, 15, 20, 19)
e13 = c(16, 17, 22, 21)
e14 = c(17, 18, 23, 22)
e15 = c(18, 19, 24, 23)
e16 = c(19, 20, 25, 24)             #Nodal indices of element 16


# We put the nodes in a list to iterate over each one of them
list_nodes=list(e1, e2, e3, e4,
                e5, e6, e7, e8,
                e9, e10, e11, e12,
                e13, e14, e15, e16)


#Global stiffness matrix
bigK <- FormGlobalKM2D(enum, list_nodes, eE, pratio,
                       edgex, edgey, height, lenscale, 0, 1)


# Global load vector
bigL <- FormGlobalLoad2D(enum, list_nodes, q, edgex, edgey)


# Specify the restrained nodes and obtain their dof
restrained_nodesABAD = HingeNodes2Dy(c(6, 11, 16, 21));
restrained_nodesBC = Symmetry2Dy(c(10, 15, 20, 25));
restrained_nodesCD = Symmetry2Dx(c(22:25));


# Combine the restrained dof
restrained_nodes = unique(c(restrained_nodesABAD, restrained_nodesBC,
                            restrained_nodesCD));


# Obtain the free dof
freenodes = ExtractFreeRows(tdof, restrained_nodes);freenodes
reducedK = FormReducedMatrix(bigK, freenodes);
reducedLoad = FormReducedLoad2D(bigL, freenodes);


unknowndisp = FindNodalDOFs(reducedK,reducedLoad);
unknowndisp[length(freenodes)]
```

Table 4.4 Effect of material length-scale parameter on the central deflection of square micro-plates

$w_c\ (a/b = 1)$		
h/l	CCCC	SFSF**
0	0.001304	0.01379
0.1	0.001247	0.01253
0.2	0.001104	0.01027
0.3	0.000928	0.00846
0.4	0.000760	0.00726

The output of the above computation is provided below.

```
> unknowndisp[length(freenodes)]
[1] 0.008461422
```

It is to be noted that a total of 59 degrees of freedom participated in the solution of the reduced matrix equation (which is stored in the variable **unknowndisp**). Therefore, the variable **unknowndisp** is a vector with 59 elements. However, only the last item (which corresponds with the central deflection of the plate) is extracted in the last line of the code for demonstration purpose. By printing out the result of the last line, it is noticed that a deflection coefficient of $0.00084614(qa^4/D)$ is produced. Although, this value is not directly reported in [14], its accuracy can be gleaned from Fig. 6 of Tsiatas [14]. As a summary, Table 4.4 highlights the trend of reduction in the deflection value for: a square plate with clamped edges (CCCC); and a square plate with a mixture of simply-supported and free edges (SFSF). The reduction in deflection value is associated with the increased stiffness in the presence of the material length-scale parameter. The values in this table are computed by changing only a few lines of the codes provided for Examples 4.3 and 4.4.

As a final note, it is pointed out that the online link for this book will contain further examples demonstrating the use of **microfiniteR** for the computations of the natural frequencies of microscale plates based on the modified couple stress theory.

4.6 Summary

In this chapter, the equations that govern the bending and free vibration behaviours of microstructure-dependent microplates are derived using the modified couple stress theory. Consistent with the other chapters, the corresponding finite element approximations of the equations are established. A necessary set of **R** functions are provided to facilitate computations of the bending response and the natural frequencies in the presence of the material lengthscale parameter (examples of natural frequency calculation are contained in the online files). Four short examples are presented to demonstrate the validity of results obtained using the package. For the different cases considered, the finite element calculations are in excellent agreement with the results reported in open literature.

References

1. K. Bhaskar, T. K. Varadan, *Plates: Theories and Applications* (Wiley, Hoboken, 2014)
2. R. Szilard, *Theories and Applications of Plate Analysis: Classical, Numerical and Engineering Methods* (Wiley, Hoboken, 2004)
3. J.N. Reddy, *Theory and Analysis of Elastic Plates and Shells* (CRC Press, Boca Raton, 2006)
4. S.P. Timoshenko, S. Woinowsky-Krieger, *Theory of Plates and Shells* (McGraw-hill, New York, 1959)
5. W. Wang, R. Lin, X. Li, D. Guo, Study of single deeply corrugated diaphragms for high-sensitivity microphones. J. Micromech. Microeng. **13**, 184 (2002)
6. H.N. Ali, I.Y. Mohammad, Modeling and simulations of thermoelastic damping in microplates. J. Micromech. Microeng. **14**, 1711 (2004)
7. R.M. Lin, W.J. Wang, Structural dynamics of microsystems—current state of research and future directions. Mech. Syst. Sig. Process. **20**, 1015–1043 (2006)
8. T.B. Jones, N.G. Nenadic, *Electromechanics and MEMS* (Cambridge University Press, Cambridge, 2013)
9. S. Ghaffari, E.J. Ng, C.H. Ahn, Y. Yang, S. Wang, V.A. Hong, T.W. Kenny, Accurate modeling of quality factor behavior of complex silicon MEMS resonators. J. Microelectromech. Syst. **24**, 276–288 (2015)
10. M. Porfiri, Vibrations of parallel arrays of electrostatically actuated microplates. J. Sound Vibr. **315**, 1071–1085 (2008)
11. J.A. Pelesko, X.Y. Chen, Electrostatic deflections of circular elastic membranes. J. Electrostat. **57**, 1–12 (2003)
12. R.C. Batra, M. Porfiri, D. Spinello, Vibrations and pull-in instabilities of microelectromechanical von Kármán elliptic plates incorporating the Casimir force. J. Sound Vib. **315**, 939–960 (2008)
13. Y.-G. Wang, W.-H. Lin, X.-M. Li, Z.-J. Feng, Bending and vibration of an electrostatically actuated circular microplate in presence of Casimir force. Appl. Math. Modell. **35**, 2348–2357 (2011)
14. G.C. Tsiatas, A new Kirchhoff plate model based on a modified couple stress theory. Int. J. Solids Struct. **46**, 2757–2764 (2009)
15. E. Jomehzadeh, H. Noori, A. Saidi, The size-dependent vibration analysis of micro-plates based on a modified couple stress theory. Physica E **43**, 877–883 (2011)
16. K.B. Mustapha, Coupled extensional-flexural vibration behaviour of a system of elastically connected functionally graded micro-scale panels. Eur. J. Comput. Mech. **24**, 34–63 (2015)
17. H. Ma, X.-L. Gao, J. Reddy, A non-classical Mindlin plate model based on a modified couple stress theory. Acta Mech. **220**, 217–235 (2011)
18. L. Yin, Q. Qian, L. Wang, W. Xia, Vibration analysis of microscale plates based on modified couple stress theory. Acta Mech. Solida Sin. **23**, 386–393 (2010)
19. L.-L. Ke, Y.-S. Wang, J. Yang, S. Kitipornchai, Free vibration of size-dependent Mindlin microplates based on the modified couple stress theory. J. Sound Vib. **331**, 94–106 (2012)
20. H. Farokhi, M.H. Ghayesh, Nonlinear mechanics of electrically actuated microplates. Int. J. Eng. Sci. **123**, 197–213 (2018)
21. B. Akgöz, Ö. Civalek, Modeling and analysis of micro-sized plates resting on elastic medium using the modified couple stress theory. Meccanica **48**, 863–873 (2013)
22. M. Asghari, Geometrically nonlinear micro-plate formulation based on the modified couple stress theory. Int. J. Eng. Sci. **51**, 292–309 (2012)

23. K. Lazopoulos, On the gradient strain elasticity theory of plates. Eur. J. Mech.-A/Solids **23**, 843–852 (2004)
24. K.A. Lazopoulos, On bending of strain gradient elastic micro-plates. Mech. Res. Commun. **36**, 777–783 (2009)
25. C.L. Dym, I.H. Shames, *Solid Mechanics: A Variational Approach*, Augmented Edition. (Springer, New York, 2013)
26. J. N. Reddy, *Energy Principles and Variational Methods in Applied Mechanics* (Wiley, Hoboken, 2002)
27. J.N. Reddy, *Energy Principles and Variational Methods in Applied Mechanics*, 2nd edn. (Wiley, Hoboken, 2002)
28. E.B. Magrab, *Vibrations of elastic systems: With applications to MEMS and NEMS*, vol. 184 (Springer Science & Business Media, 2012)
29. S.S. Rao, *The Finite Element Method in Engineering* (Elsevier Science, Amsterdam, 2010)
30. M.A. Bhatti, *Advanced Topics in Finite Element Analysis of Structures: With Mathematica and MATLAB Computations* (Wiley, Hoboken, 2006)
31. M. Petyt, *Introduction to Finite Element Vibration Analysis* (Cambridge University Press, Cambridge, 1998)
32. J.N. Reddy, *Introduction to the Finite Element Method* (McGraw-Hill, New York, 1993)
33. R.D. Cook, W.C. Young, *Advanced Mechanics of Materials* (Prentice Hall, New Jersey, 1999)

Printed in the United States
By Bookmasters